D1327366

CHERNOBYL
A TECHNICAL APPRAISAL

Proceedings of the seminar organized by the
British Nuclear Energy Society held in London
on 3 October 1986

British Nuclear Energy Society, London

Conference sponsored by the British Nuclear Energy Society and the Institution of
Civil Engineers

First published 1987

British Library Cataloguing in Publication Data
Chernobyl: a technical appraisal:
    proceedings of the seminar organized
    by the British Nuclear Energy Society
    held in London, 3 October 1986.
    1. Nuclear power plants — Ukraine —
    Chernobyl — Accidents
    I. British Nuclear Energy Society
    363.1'79   TK1362.S65

ISBN: 0 7277 0394 3

Published for the British Nuclear Energy Society by Thomas Telford Ltd, Telford House,
1 Heron Quay, London E14 9XF

Printed and bound in Great Britain by Robert Hartnoll (1985) Ltd, Bodmin, Cornwall

CONTENTS

Opening address

J. RIMINGTON, *Director General, Health and Safety Executive*

It can be useful for an outsider such as myself to make a few
banal points to the experts which are already obvious, but
which can be lost to view once the plunge into deeper waters
begins. The principal immediate effect of the Chernobyl
accident is political, and throws the future of nuclear power
generation into the political cauldron. There is a tendency
nowadays to be very suspicious of events. I think it began
with the irritating determinist argument which claimed that
things cannot happen otherwise than as they do, and that
therefore events are unimportant.

Whatever damage the determinist argument failed to do to
events,William Randolph Hearst completed by his revelation of
the media's power to create events – a power that has since
impressed itself more and more, not least in the nuclear
sphere. But so far Chernobyl defies all this. It is an
event capable of changing a whole direction of affairs.
Clearly Chernobyl was not invented. No one pretends that it
did not actually happen. Nobody has yet supposed that the KGB
may have decided to devastate large tracts of the Ukraine to
abash the West by subsequent demonstration of candid
behaviour.

Chernobyl has gripped the popular mind. Today the issue of
nuclear power is the most important issue in Labour Party and
Trade Union politics. It, and the largely false association
with the bomb, has riven the Alliance parties. There are
indications from many other parts of Europe and over the
Atlantic that similar political interest has been generated on
the basis of popular reaction.

Thinking about nuclear power is no longer a fringe activity.
For the first time, ordinary people are thinking about its
future, and since the solid mass of people are thinking, the
debate is likely to be long and uncertain. What questions are
they asking? I suggest two.

First is there a risk which, no matter what the
consequences, should not be taken? Second (shades of
determinism), was it bound to have happened, taking account of
human nature, sometime, somewhere, and is it going to happen
again? I ask that because I do not think that people think
familiarly in terms of factors of risk.

At the end of the day scientific opinion has to satisfy itself as to what the causes of Chernobyl were. Those concerned with safety and above all the regulatory authorities must see that those causes, most of all the general ones, are addressed. They must convince people, if nuclear power deserves to continue, that they can and will see that the lessons are applied and that they will be effective. Chernobyl was an event after all, not a premonition.

In one sense Chernobyl was a case of human error, but the error has a general tendency. It sprang from the eternal tension between efficient performance and safety. These poles must always exist and safety must be paramount. Why was the balance allowed to swing so decisively to one pole on that particular occasion at Chernobyl? What are the lessons to be learnt for safety regulations, both by the regulator and within the power generating organization? The designers were well aware of the natural instabilities of this particular design which fact strikes me as having in its basic concept the powerful simplicity of Newcomen's pump. Any nuclear physicist would have been aware of the nature of the modifications taking place in the physics of the core as the control rods advanced and withdrew their input, but it would appear that the entire experiment was in the hands of the engineers. What does that say about the structure of safety regulation within and beyond the Russian generating organization? Do the engineers talk to the physicists? Does the safety regulator have the necessary presence, power and independence?

The ergonomists have important questions to answer, and they concern the kind of complex situation best controllable by the human mind. They also concern the balancing of the two poles, generation and safety. The most fundamental principle in the British attitude to safety precaution is that the engineered safeguard is to be preferred in the vast majority of situations.

Design and control characteristics of the RBMK reactor

D. R. SMITH, D. P. LUCKHURST AND A. R. MACCABEE,
National Nuclear Corporation Ltd

SYNOPSIS

This paper describes Chernobyl 4 Nuclear Power Station, highlighting those features of the design which had some bearing on the accident there on 26th April 1986 and on the course of events leading up to the accident.The general layout of the plant, the main safety systems,and the core physics characteristics are described. The control and protection of the reactor are discussed, emphasising the reliance placed upon the plant operators to maintain the reactor within safe limits.

INTRODUCTION
1. The Chernobyl plant is a graphite moderated channel tube reactor to which the Russians have given the acronym RBMK. The combination of a graphite moderator with a pressure tube coolant circuit in a commercial nuclear power station is unique, although its parentage can be traced to the early reactors built to produce military plutonium.

2. In the early 1970s the Russians embarked upon an ambitious power programme aimed at commissioning some 15,000 MW(e) of plant by the end of 1980. The PWR and the RBMK reactor types were the candidates for this programme. Four small PWRs were operating in Novo Voronezh and two small RBMK reactors at Byeloyarsk. The RBMK was chosen to provide the larger share of the nuclear programme primarily because its smaller components were easier to manufacture and transport to site in Russia.

3. The chief design features of the RBMK reactors are:
(a) Vertical fuel channels with onload refuelling.
(b) An 18 pin cluster, each pin comprised of uranium dioxide fuel pellets clad in zirconium alloy.
(c) A graphite moderator between the fuel channels.
(d) A boiling light water coolant with direct steam supplied to the turbines.

4. The feasibility of incorporating steam superheat channels into the RBMK has been demonstrated at Byeloyarsk, but this has never been followed up in the commercial plants.

5. At the IAEA Conference held in Vienna in August 1986 the Russians acknowledged that several fundamental shortcomings in the RBMK were recognised in the early 1970s. They considered that these shortcomings could be accommodated by suitable provisions in the detail design, and that they were outweighed by the advantages for manufacture and transport referred to earlier. The events at Chernobyl demonstrate the implementation of the detail design provisions to have been totally inadequate. The fundamental design shortcomings listed by the Russians were the following:

(a) A positive void coefficient of reactivity in the coolant.

(b) An unstable core power distribution requiring a complex control system.

(c) A complex coolant circuit.

(d) A large amount of stored energy in the metal and graphite structure.

6. Prior to the Chernobyl accident the history of the RBMK had been very successful. The first 1,000 MW(e) RBMK was commissioned at Leningrad in 1974 and this was quickly followed by further units. The Leningrad, Kursk and Chernobyl power plants each have 4 units built in pairs, each unit supplying twin 500 MW(e) turbine generators. The first 2 (of 4) units are operating at Smolensk and 2 more are being constructed at Chernobyl. The first of 2 larger 1500 MW(e) versions of this reactor was put into operation at Ignalino in Lithuania in 1984. The physical size of this reactor is similar to that of the 1,000 MW(e) RBMK but it has a 50% higher power density.

7. The operating RBMKs achieved an average load factor of about 80% during 1985. In safety terms there have been practical demonstrations of the ability of the design to sustain significant faults. For example, at Kursk in January 1980 a station black-out occurred which was sustained satisfactorily. There has also been a number of cases of feedwater transients, none of which presented a serious safety challenge to the plants.

8. The UK nuclear industry first became involved with the RBMK reactor in 1975, as part of a policy of improving Anglo-Soviet relations. A delegation from the British Nuclear Forum visited Russia in the Autumn of that year and this was followed by a return visit by the Russians. In 1976 similar visits took place under an AEA collaborative agreement with their Russian counterparts on Pressure Tube

Reactors. The main objective of these visits was to see what could be learnt from the very significant progress that the Russians were making in the design and construction of the RBMK, at a time when the UK was engaged in the design of a commercial version of the Winfrith steam generating heavy water reactor, which is also a direct cycle pressure tube reactor.

9. NNC participated in both these visits, and undertook a study to compare the RBMK and SGHWR with a view to identifying areas where detailed exchanges of information would be beneficial. A report on the NNC study, issued in March 1976, concluded that various features affecting safety would make it very difficult to transplant the Russian concepts into the UK. The main concerns highlighted in the Report were the following:

(a) Safety implications of positive reactivity coolant void coefficient need to be checked.
(b) Control rod investment inadequate by UK standards.
(c) Positive graphite temperature coefficient creates stability problems.
(d) No secondary shutdown system provided as required in the UK.
(e) No protection against rupture of large pipes in reactor coolant circuit.
(f) No protection against coolant stagnation during LOCA.
(g) Fuel channel tube rupture likely to propagate through the graphite structure.
(h) Graphite temperatures probably unacceptably high.
(i) Coolant circuit more vulnerable than integral pressure circuits.
(j) No containment building provided.

10. The study was limited by a lack of detailed information on the RBMK, but it is noteworthy that a number of the NNC reservations coincide with the shortcomings subsequently acknowledged by the Russians in the recent Vienna conference. Other NNC reservations have been mitigated by design changes incorporated in the later Russian plants, including Chernobyl, notably the provision of emergency core cooling and containment systems to protect against coolant pipe ruptures. However, most of the reservations remain, and several contributed directly to the accident at Chernobyl. The one comment which has proved unjustified is that relating to control rod investment. This comment assumed that boron steel rods were used whereas it later emerged that the RBMK employs boron carbide which is a more effective absorber of neutrons.

11. In the following sections of this session the salient features of the RBMK design are described with comments regarding their adequacy against UK safety standards and practices.

OVERVIEW OF THE RBMK

12.   The main features of the RBMK are shown in Fig.1.   At
the centre of the unit is the reactor core with its
supporting structures and biological shielding.   The reactor
coolant ciruit, consisting of a complex array of pipework
and valves, supplies water to the fuel channels and removes
steam/water mixture to the steam drums.   Above the reactor
is the reactor hall, containing the fuelling machine.   A
containment building surrounds the reactor and primary
circuit.

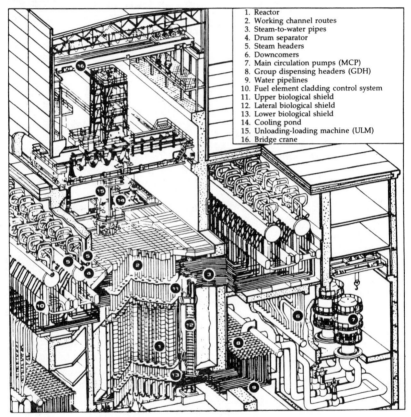

1. Reactor
2. Working channel routes
3. Steam-to-water pipes
4. Drum separator
5. Steam headers
6. Downcomers
7. Main circulation pumps (MCP)
8. Group dispensing headers (GDH)
9. Water pipelines
10. Fuel element cladding control system
11. Upper biological shield
12. Lateral biological shield
13. Lower biological shield
14. Cooling pond
15. Unloading-loading machine (ULM)
16. Bridge crane

Fig. 1.   Sectional view of RBMK 1000 reactor

REACTOR AND SUPPORTING STRUCTURES

13.   The reactor and its supporting structures are shown
in Fig.2.   The reactor core is cylindrical in shape, with a
diameter of 11.8 m and a height of 7m.
It consists of graphite blocks assembled into columns, with
vertical, cylindrical openings which house a total of 1661
fuel channels and 211 control channels.

14. The whole of the reactor is located in a leaktight cavity formed by the upper and lower shields and a cylindrical steel shell. The cavity is filled with helium-nitrogen mixture (85-90% He, 15-10% N_2) which inhibits oxidation of the graphite and improves heat transfer from the graphite to the cooling channels. The He/N_2 mixture is circulated through a clean-up system where it is monitored for temperature and moisture content.

15. Outside the reactor cavity is a lateral shield which takes the form of a cylindrical reservoir (annular in cross-section) 19m outside diameter and 16.6m inside diameter.The tank is fabricated from low alloy steel, and is divided internally into 16 vertical leak-tight compartments filled with water. The tank is supported from below by the building structure.

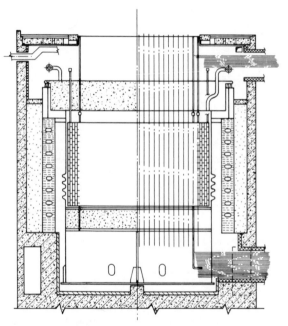

Fig. 2. Reactor cavity structure

16. The upper biological shield is a fabricated steel structure,17m in diameter and 3m in height. It is penetrated by standpipes which accommodate the fuel and control channel assemblies. The space between the standpipes is filled with serpentine concrete. The upper shield supports the loaded fuel channels, the floor of the reactor hall above, and the reactor coolant outlet piping. This entire structure is in turn supported on roller-type supports by the lateral shield tank.

17. The lower shield is similar in design to the upper structure, but 14.5m in diameter and 2m high. The structure supports the graphite stack and the reactor coolant inlet piping, and is penetrated by steel tubes which accommodate the lower ends of the fuel and control channels. The lower shield is supported from below, by a fabricated steel structure, which transmits the weight of the graphite stack to the building.

18. Above the upper shield is a gallery containing the upper ends of the standpipes and the reactor outlet (or riser) pipes. Above this are slabs which form biological shielding for the reactor hall, and act as heat insulation. The slabs rest on the standpipes and can be removed for access to the standpipe for refuelling.

19. The fuel channel itself consists primarily of zirconium alloy (Zr, 2.5% Nb) tube, 88mm external diameter with a wall thickness of 4mm. Steel/zirconium alloy transition pieces facilitate the attachment of upper and lower steel end-pieces. The upper end piece is rigidly attached to the steel standpipe which penetrates and is sealed to the upper shield structure. The lower end piece is sealed to the lower shield structure through a bellows unit which compensates for the difference in thermal expansion between the fuel channel and the reactor cavity structure.

20. The gap between the fuel channel and the graphite is small, so that failure of the pressure tube would result in significant pressure loads on the graphite blocks. This would tend to lift the bricks and would be likely to damage them, particaularly after irradiation. Consequently the graphite stack appears to be vulnerable to pressure tube failure, and there is a likelihood of propagation to other tubes.

21. Taken together with the relatively low design pressure of the reactor cavity and the lack of rigid connection between the surrounding structures, these features would be unlikely to be acceptable in the UK.

PRIMARY COOLANT CIRCUIT
22. The primary circuit (illustrated in Fig.3) supplies water to the fuel channels and removes the steam/water mixture, which forms as a result of the heat taken up from the fuel assemblies, for subsequent separation in the steam drum. It consists of two loops, similar in their arrangement and equipment, which function in parallel and independently; each loop removes heat from half of the reactor's fuel assemblies.

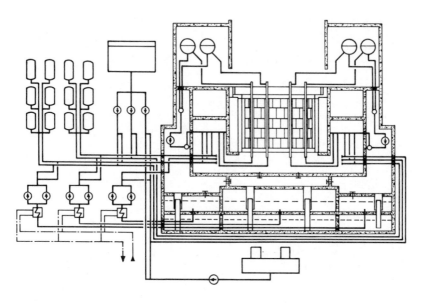

Fig. 3.   Reactor cooling circuit and emergency core cooling
systems

23.   The pressure tube concept employed in the RBMK
reactors results in a much more complex pressure circuit
than has become customary in the UK.   In each half of the
coolant circuit, water from the suction header passes
through four pipes to the main circulating pumps.   Under
normal operating conditions at normal power three of the
four main circulating pumps are in operation, with one held
in reserve.   Water leaves the main circulating pumps at a
temperature of 270°C at a pressure of 8.4 MPa (1218 psi)
through pressure pipes, in each of which are installed in
sequence a non-return valve, an isolation valve and a
throttle valve.   Coolant then flows into the main
circulating pump pressure header, from where it passes into
the 22 distributing headers, which have non-return valves at
their inlets.   Individual lines lead into the fuel channels.
The flow rate through each channel is controlled by means of
isolating and regulating valves.   As it passes through the
fuel channels, the water surrounding the fuel elements is
heated to saturation temperature, partially evaporates
(14.5% on average) and the steam-water mixture at a
temperature of 284.5°C and a pressure of 7 MPa (1015 psi)
flows through the individual riser pipes into the steam
drums where it is separated into steam and water.

24.   The water which has been separated out is mixed at
the drum outlets with feed water, and flows through 12
downpipes (from each drum) into the suction header.   The
temperature of the water flowing into the suction header
depends on the rate of steam production of the reactor.
When this decreases, the temperature increases somewhat
because of the changing ratio of water from the drum
separators, at a temperature of 284°C, and feed water, at
a temperature of 165°C. Consequently, when the reactor
power is below 500 MW (t) during start-up or shut down
operations, the flow rate through the primary circuit is
controlled using throttle-type control valves to reduce the
flow from the normal rate of 8000 m^3/hr per pump, to the
range 6000-7000 m^3/hr.   This is necessary to ensure that
the temperature at the main circulating pump inlet maintains
the necessary cavitation margin, and to maintain steam
production in the core.

25.   The reactor would be tripped on indications of high
or low steam drum level, high steam drum pressure, low
feedwater flow or loss of more than two main coolant pumps
in either half of the reactor.   These trips are known to
have been blocked by the operator prior to the accident, in
a way which would be impossible in the UK plant.

EMERGENCY CORE COOLING SYSTEM (ECCS)
26.   The ECCS is designed to maintain core cooling in the
event of pipe failures in the main coolant system and in
some "intact circuit" faults in which the supply of water to
the fuel channnels is interrupted for other reasons, [for
example: loss of feedwater or loss of power supplies.]
Actuation of the ECCS is initiated automatically, but it is
known that the operators blocked ECCS actuation, prior to
the accident.   Because the reactor coolant system is
arranged in two separate, independent loops, the ECCS must
be arranged to meet the differing functional requirements
associated with the breached and intact halves of the
reactor. The Chernobyl ECCS consists of three sub-systems
(see fig.3) all connected to the distribution headers of the
coolant system.   One provides flow to the damaged half only,
during the early stages (up to 100 sec) of a breach. During
this time cooling of the undamaged half is via the normal
RCS route.   The other two ECCS sub-systems provide long term
cooling for both halves of the reactor.

27.   Short term cooling of the breached half is provided
by three groups of ECCS equipment, comprising an accumulator
system and a pumped system. Two groups consist of six
accumulator tanks, containing water with a nitrogen gas
blanket, maintained at 10 MPa (1450 psi).   Each group is
capable of delivering 50% of the maximum flow requirement to
the damaged half of the reactor, for not less than 100
seconds after the initiation of the breach. Fast acting

valves are used to initiate ECCS flow within 3.5 seconds of the break in normal water supply to the reactor. The third group utilises the normal feed pumps,which are electrically driven and which draw water from the deaerators. The pumps are realigned to feed directly to the ECCS headers. This route is capable of delivering 50% of the maximum flow requirement to the damaged half of the reactor.

28. Long term cooling of the damaged half of the reactor is provided by three groups of ECCS equipment. Each group consists of two pumps connected in parallel and is capable of delivering 50% of the maximum flow. The pumps draw water from the pressure suppression pools beneath the reactor, the water being cooled by service water in heat exchangers in the pump suction lines.

29. Long term cooling of the intact half of the reactor is provided by three groups of ECCS equipment. Each group consists of a single pump drawing water from a condensate storage tank, and is capable of delivering 50% of the required flow.

30. It is a requirement that the ECCS must fulfil its function in the event of a loss of normal power supplies coincident with the design basis accident. The normal power supplies for pumps and valves in the ECCS are derived from the grid,(see para 39) but in the event that the grid is lost then power can be obtained from the running turbo-generator(s). During turbine run-down, the turbo-generator continues to support the electrical system for a limited period (45-50 seconds in the case of the feedpumps which contribute to short term ECCS duty). Subsequently, diesel generators are brought in, to power the pumps in those sub-systems which contribute to longer term cooling. ECCS valves which cannot accept interruption in supplies are supported by batteries.

31. The emergency systems provide 150% redundancy. The design is claimed to take account of unproductive loss of coolant through the breach, and a single active or passive failure within or outside the ECCS. Unavailability of part of the system, due to maintenance, does not appear to be taken into account. Whilst a detailed assessment has not been made, the complexity of the system and its dependence upon valve alignment would probably result in lower reliability than would be required in the UK.

CONTAINMENT SYSTEMS

32. The Chernobyl building structure is designed to contain and confine the release of radioactivity following failures in certain parts of the reactor coolant circuit. The containment structure and associated systems mitigate the effects of pressure part failures in the lower parts of

the downcomer, the pumps, the pressure headers, distribution headers and inlet pipework. However, failures in the upper parts of the fuel channel, the riser pipes, the steam separators themselves and the upper portions of the downcomers are not catered for. This can be seen in Fig.3, which shows the boundary of the containment structures together with the primary circuit.

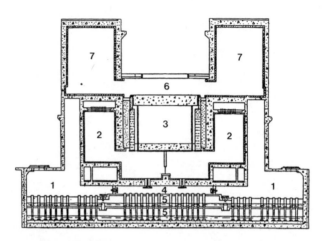

1 Pump compartments
2 Group header compartments
3 Steel reactor vault
4 Surface condenser tunnel

5 Suppression pools
6 Riser pipe gallery ⎫
7 Steam separator ⎬ Secondary
   compartment ⎭ containment

Fig. 4. Primary and secondary containment layout

33. The principle features of the containment system are shown in Fig.4.
(a) leaktight compartments enclosing the pumps (1) designed for an overpressure of 0.45 MPa (65 psi).
(b) compartments enclosing the distribution headers and reactor inlet piping (2) which will tolerate an overpressure of 0.08 MPa (11.6 psi) and which are vented to the leak tight compartments via non-return valves.
(c) the reactor cavity which will tolerate an overpressure of 0.18 MPa (26psi) and is vented to the leaktight compartments.
(d) within the leaktight compartments are a steam distribution corridor (4) and a two-storey pressure suppression pool.(5), partially filled with water. The internal partitions are penetrated by non return valves and venting channels.
(e) the riser pipe gallery (6) and the steam drum compartment (7), which are not designed to withstand overpressure and are normally maintained at a pressure slightly below that in the reactor hall.

34. Following pressure circuit failures in the pump compartments or the group header compartment steam/water and air are vented to the suppresssion pools via non-return valves and venting channels. The results of calculations presented by the Russians show that the compartment pressures remain below the design values for the cases of failure of a pump pressure header, a distribution header or a single fuel channel within the reactor cavity.

35. There is provision for heat removal from the containment compartments and from the pressure suppression pools. Surface condensers in the steam corridor remove heat only during accident conditions. A sprinkler system operates both during normal operation and during accidents. Water is drawn from the suppression pool and is cooled in a service water heat exchanger. The water is then directed to the air space above the suppression pool, where it mixes and cools the air, and to jet coolers in the upper (hottest) part of the containment compartments. The jet coolers entrain air, thereby cooling the environment and removing radioactive aerosols and steam. The water is collected and returned to the suppression pool.

36. Provision is made to maintain the concentration of hydrogen below 0.2% (by volume). Hydrogen is present in the containment volume during normal operation owing to coolant leakage (assumed to be at a rate of up to 2 t/hr). During accident conditions hydrogen may also arise from zirconium-water reaction. To cater for this, air is drawn from the containment volume at the rate of 800 m^3/hr during normal operation, and is discharged to atmosphere via filtration plant. The purge is automatically discontinued immediately following a failure of the coolant circuit and is then reinstated manually after a period of 2-3 hours, as hydrogen accumulates.

37. It is generally accepted in the Western World that water-cooled power reactors must have three barriers to release of radioactivity. The first two barriers (the fuel cladding and the primary circuit) were present at Chernobyl, but the third (the building) was only capable of containing activity following failures in part of the circuit. This must be viewed as a deficiency against UK standards.

POWER SUPPLY SYSTEMS
38. The Chernobyl 4 power systems are illustrated in Figs. 5 & 6. The station has two main generators connected to the 750 kV grid via a single generator transformer. Two generator switches are installed in series with each main generator and a connection for a unit transformer is made between each pair of generator switches. The plant also has a 330kV connection to the grid via a station transformer, although this route is normally isolated from the main electrical system.

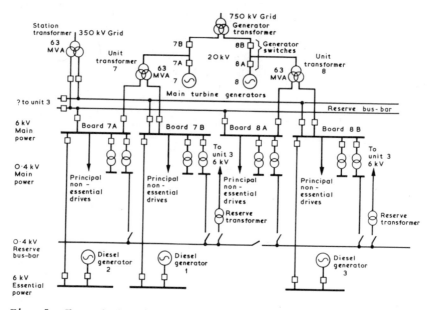

Fig. 5. Chernobyl unit 4 main electrical system

39. Normal power supplies for house loads are obtained via the unit transformer. Supplies can also be derived from the following sources in various operating modes:

(a) following a reactor or turbine trip, from the 750kV grid via the generator transformer

(b) following loss of the 750kV grid or generator transformer from the 330kV grid via the station transformer.

(c) Following faults initiated by reactor coolant circuit failure with co-incident loss of normal power supplies and turbine trip, the running down turbo-generators can be used to support the electrical system loads for a limited time. This mode of operation is used to support the feed pump in its ECCS mode (see para 30) for a period of 45 - 50 seconds.

(d) following loss of all off-site supplies, from diesel generators

40. Following a loss of off-site power, any running essential loads are tripped, the diesel generators are started automatically and connected to the electrical system within 15 seconds. Essential drives required by the incident are sequence started from the available power source (grid or diesel generator). Auxiliaries which cannot

accept the interruption of supplies are supported by batteries. The three groups of essential safety equipment and their controls have independent power supplies.

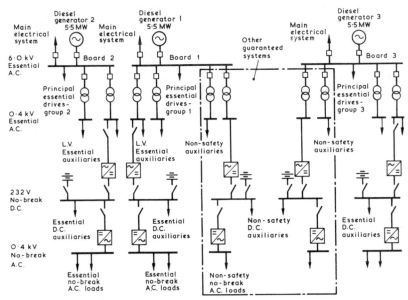

Fig. 6. Chernobyl essential electrical system

SUMMARY OF PLANT DESIGN.

41. The main features of the Chernobyl 4 Nuclear Power Station having some influence on the accident there on 26th April 1986, are as follows:

a. the reactor cavity is formed by a relatively thin cylindrical shell with modest pressure retaining capability (0.18 MPa, 26psi, abs) to accomodate failure of a single fuel channel.

b. Whilst the reactor cavity is capable of tolerating pressure tube failure, it is not clear that the graphite stack is capable of withstanding the disruptive effects of channel rupture.

c. The structures immediately surrounding the reactor cavity are massive but are loosely connected to one another.

d. The building structure above the reactor is not designed to withstand internal pressure. The building structures around and below the reactor are designed for a pressure duty of 0.45 MPa (65 psi). They consist of a number of separate compartments interconnected by vent pipes and non-return valves.

e. The electrical power systems include the facility to use a running-down turbo-generator to support auxiliary systems. This mode of operation is of some

15

importance in achieving the required flow from part of the ECCS in the event of loss of normal power supplies.
f. The RBMK design includes provision for reactor trip in the event that the plant goes outside defined limits of operation, and includes fast-acting systems for emerging core cooling. The systems are actuated automatically but the operators are able to veto actuation.

PHYSICS CHARACTERISTICS OF THE REACTOR

42. The physics characteristics of the RBMK reactors are determined from the geometry and materials of the fuelled lattice cell in figure 7, which combine certain features of both graphite and water moderated systems.

43. The 18 fuel pins are arranged in two rings of 6 and 12 pins around a central tie rod, supported by grids at 36cm intervals, within a pressure tube of 8cm internal diameter. The fuel pins are cooled by water flowing within the pressure tube, from bottom to top of the fuel assembly length. The pressure tube is positioned within a graphite block, separated from the main bulk of the block by layers of machined graphite annuli designed to make firm contact alternately with the graphite block and the pressure tube wall, to improve radial conduction of heat from the graphite into the water coolant.

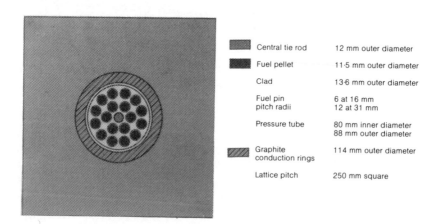

Central tie rod	12 mm outer diameter
Fuel pellet	11·5 mm outer diameter
Clad	13·6 mm outer diameter
Fuel pin pitch radii	6 at 16 mm 12 at 31 mm
Pressure tube	80 mm inner diameter 88 mm outer diameter
Graphite conduction rings	114 mm outer diameter
Lattice pitch	250 mm square

Fig. 7. Section through fuelled lattice cell

Table 1. Comparison of the Chernobyl RBMK moderating ratio with typical AGR and PWR values.

	Moderating ratio		Vmod/Vuo2
	Chernobyl RBMK	AGR	PWR
Graphite	29	26	–
Water	1.5	–	1.6

44. A comparison of the moderating ratio by volume of the RBMK, AGR and PWR reactors in Table 1 gives an immediate insight into the dominant reactor characteristic. The moderating ratio by volume derived from the graphite alone is similar to the AGR and, although the graphite density in the RBMK is approximately 10% lower than in the AGR, the RBMK fuelled lattice is still very close to the peak of the curve of the infinite neutron multiplication constant, K_∞ , against moderating ratio, without the cooling water present. In these circumstances, the neutron absorption in the cooling water is increasingly important relative to its moderating properties, so that the addition of the coolant , itself similar in volume to the PWR moderating ratio, leads to a design which is significantly overmoderated and at the same time has a high thermal absorption in the material whose presence, by voiding, connot be guaranteed.

45. Partial voiding of the water coolant from operating conditions can therefore be expected to increase K_∞ by reduction in moderation and by reduction in thermal absorption of the system.

46. Typical supercell results, allowing for control rod vacancies are given in Table 2. In the reactor, the neutron leakage has a stabilising effect, so that the increase in reactivity of the core due to coolant voiding is reduced. Nevertheless, the RBMK 1000 core is large, 7 metres in height and 11.8 metres in diameter, so that the geometric buckling is small and the void coefficient of the core at equilibrium fuel cycle, full power conditions, with all control rods withdrawn, is positive over the entire range of coolant voidage.

TABLE 2
Equilibrium, full power, all rods withdrawn.
average void coefficients over coolant voidage range

Water/Steam mixture density range(gm/cc)	0.75 – 0.55	0.55 – 0.25	0.25 – 0.01
Void Coefficient $\Delta K/K$ per % void.	$2_{10^{-4}}$	$2_{10^{-4}}$	$3_{10^{-4}}$

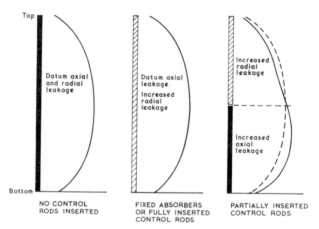

Fig. 8. Reactor leakage changes due to absorbers and control rods

47. As mentioned previously, neutron leakage has a stabilising effect, making the void coefficient less positive. The neutron leakage within the reactor can be increased by the introduction of control rods or fixed absorbers, as depicted in figure 8. With full-length fixed absorbers or fully inserted control rods in the reactor, there is little change in the axial flux shape, but an increase in radial neutron leakage into the absorber sites. With partially inserted control rods, the axial flux shape is modified, so that the dominant changes are increased radial leakage in the plane of the inserted rods and increased axial leakage below the plane of the inserted rods. Calculations of the reactor void coefficient with partially inserted rods are not available, but the results of xy supercell calculations representing the equivalent of full insertion of half the total control rods, in Table 3, are adequate to demonstrate the scale of their effect.

Table 3.Effect of control rods and absorbers on void coefficient.

Water/steam mixture density range(gm/cc)	0.75-0.55	0.55-0.25	0.25-0.01
EQUILIBRIUM			
No rods inserted	2×10^{-4}	2×10^{-4}	3×10^{-4}
50% of rods inserted	0	0	0
INITIAL CHARGE			
241 fixed absorbers	-1×10^{-4}	-2×10^{-4}	-3×10^{-4}

48. The presence of Pu 239 at equilibrium conditions makes a positive contribution to the void coefficient, so the value for the unirradiated initial charge is expected to be less positive. In addition, the initial charge has a high built-in reactivity, offset by the replacement of 241 fuel assemblies by fixed absorber assemblies. The additional leakage into the absorber channels and the unirradiated fuel ensure that the void coefficient of the initial charge is negative, without inserted control rods.

49. At the time of the accident, Chernobyl 4 was part way through the approach to equilibrium and contained 1659 fuel assemblies, 1 absorber assembly and 1 empty channel. Since the average fuel burn-up at the time was 10.3 GWD/Te and 75% of the fuel was between 12 and 15 GWD/Te, compared with a design discharge irradiation of 20 GWD/Te, it can reasonably be assumed that the void coefficient was approximately at the equilibrium value and further discussion will concentrate on the equilibrium condition.

50. In a design such as the PWR, the water in the primary circuit flowing through the core is maintained with a subcooled margin, so that voidage in normal operation is minimal and the void coefficient only has relevance to fault conditions. The RBMK reactors however are of of the boiling water type, so that voidage occurs in the primary circuit, the water entering an average fuel channel with a small subcooled margin but leaving the channel as a water-steam mixture of approximately 15% quality.

51. The positive void coefficient of the RBMK is therefore an important parameter in both normal and fault conditions, since an increase in core reactivity results from any reduction in coolant flow or increase in coolant inlet temperature.

52. A positive value of the void coefficient does not itself mean that a reactor design is unsafe.More important considerations are the magnitude of the positive void coefficient, determining the reactivity insertion as the

coolant voidage increases, and the short term effect of the subsequent power increase. The latter is described by the power coefficient, which combines the reactivity change due to the increasing fuel temperature and coolant voidage as the power increases. The fuel temperature (or Doppler) coefficient is negative, typically -1.2_{10} $^{-5}$ per ^{o}C at full power, equilibrium fuel cycle conditions, so the power coefficient could be negative or positive depending on the magnitude of the positive void coefficient. A negative value of the power coefficient means that a power increase can be retarded or limited by the associated negative reactivity feedback, at least allowing time for corrective action or reactor trip, but a positive power coefficient leads to an acceleration of the initial power excursion by adding more reactivity. The severity of the power transient is then dependent on the detailed balance of the positive feedback and reactivity taken up by control rod movement or reactor trip as a function of time.

53. The Russian statement provided no detailed information on the possible range of variation of the power coefficient with reactor conditions.Recent calculations carried out at Berkeley Nuclear Laboratories, in Table 4, suggest that, at equilibrium fuel cycle conditions, the power coefficient is negative at full power over the full range of rod insertions, but at low power it can be positive at low rod insertion and very dependent on the coolant flow rate.

Table 4. Variation of the power coefficient in 10^{-5} per % of nominal full power at equilibrium fuel cycle.

Power % Flow %	100 100	7 87	7 21
Unrodded	−2.5	+13	+38
Rods inserted	−6.4	+3	+21

54. These compare with the Russian value of -1.6_{10} $^{-5}$ per percent power at the "working" point,but it is not known whether this is a best estimate or conservative value, or whether the 'working' point implies full power conditions or the worst value over a permitted range of operating conditions.

55. The temperature coefficient of the graphite at equilibrium fuel cycle, full power conditions was quoted as 6_{10} $^{-5}$ per ^{o}C, but the relatively long time constant of the graphite temperature response to power changes limits the importance of this coefficient in analysis of rapid transients.

OPERATING REGIME AND SAFETY ANALYSIS

56. The safety design philosophy of the RBMK reactors has not been provided in any detail. It is possible, however, to identify a number of requirements from the characteristics of the reactor core and to compare these requirements with the known actions of the protection system and the limited information available from the accident analysis on breaches of operating regulations.

57. From the preceding section, the operating regime should be limited so that:
1. The void coefficient can be positive but limited in magnitude.
2. The power coefficient cannot be positive and should be significantly negative.

58. Both factors are affected by control rod worth and insertion and power level, so the operating regime should be defined for safety analysis as limitations on these two parameters. Establishing a safety case in terms of simply expressed bounding requirements on measurable parameters can involve much more extensive analysis than the alternative of building in a limited range of reactor characteristics at the design stage.

59. A number of important considerations in the definition of the bounding requirements can be identified and compared with what is known of the RBMK operating regulations.
These are:

a. the worth of control rods inserted during normal power operation should be sufficient to restrict the reactor void coefficient to a level consistent with a negative power coefficient. The RBMK has an on-line computation of the 'operative reactivity margin', quoted as 1% worth in inserted rods.

b. normal power operation should be permitted only above a lower power limit and operation below this level should be subject to special restrictions. In particular the physics characteristics of the core indicate that additional inserted control rod worth may be required in startup which would limit the availability of the plant after reactor trip. Although not included in the Russian report, Chernobyl had a lower limit of 20% of full power in normal operation, and startup was precluded between 1 and 24 hours after trip from full power.

c. since any coolant flow variation or fault can inject significant amounts of reactivity into the core, the subsequent controlling or trip action should be immediately effective and sustainable. The control rods in normal operation should therefore be maintained within an insertion range of high differential worth and any trip should lead to a rapid or sustained high reactivity take-up. There is no record of any rod insertion limits in normal operation and a reactor trip causes all rods to motor into the core at a speed of 0.4 metres per second. The RBMK safety analysis did, however, identify a need for an initial trip worth of β per second, where β is the reactivity equivalent of the delayed neutron fraction, about 0.55% at equilibrium fuel cycle. This requirement was presumably dominated by LOCA analysis and implies partial insertion of large numbers of rods in normal operation.

d. tripping rods into the core not only takes up reactivity but also makes the void and power coefficients less positive or more negative. In some circumstances, this stabilising contribution may eliminate the initial cause of the trip, allowing recovery to a stable operating condition if the rods could be stopped. The RBMK protection system design, includes a level which allows the rods to stop if the trip condition clears.

60.   The following section describes how the reactor is operated and protected, identifying the role of the operator.

REACTOR OPERATION AND PROTECTION

61.   The Chernobyl design incorporates an in-core flux monitoring system, which is claimed to be essential for economically effective and safe operation. The system consists of 130 single detectors distributed in the xy plane at a fixed axial position and 12 strings of 7 detectors, again distributed in the xy plane.

62.   Readings from the in-core detectors are processed in a centralised monitoring system, known as SKALA, which combines the detector readings with reactor calculations provided periodically by links to an external computer to derive the integrated power of each fuel assembly and the associated axial power profile. These power distributions are combined with measured assembly flow rates and inlet conditions to calculate the margin to critical heat flux in each assembly, to ensure no departure from nucleate boiling.

63.   Unacceptably low margins to critical heat flux in any assembly are displayed to the operator on a reactor mimic plan every 5 to 10 minutes, together with a recommendation on flow adjustment to the assembly, while the automatic control system initiates a local or reactor power setback at zero margin.

64.   The flow to each assembly can be adjusted by the operator and in equilibrium fuel cycle an adjustment is made

at least twice in the life of each of the 1661 fuel
assemblies, to optimise the reactor output.

65. Stability of the reactor power and power
distribution in normal operation is controlled automatically
by two discrete groups of rods:

1. the automatic mean power regulating group, either of
    two groups of four, controlling reactor power in
    response to signals from detectors in the lateral water
    shield outside the core

2. the local automatic control group of rods divided into
    12 single rod zones, controlling instabilities in the
    radial azimuthal power distribution, in response to
    signals from local detectors. This control function
    is backed up by a local emergency protection system,
    which either freezes the automatic rod withdrawal or
    inserts local rods.

66. The operator is expected to trim the radial azimuthal
power distribution by movement of any of 142 manual rods to
optimise power output and also to eliminate the first
harmonic in any axial instability by movement of manual rods
into the top of the core or 21 shortened absorber rods into
the bottom of the core, as appropriate.

67. Although not confirmed, it is believed that the
operator is required to maintain the mean power regulating
rods and the local automatic control rods within a specified
insertion band by movement of the manual rods.

68. The SKALA monitoring system computes the reactivity
worth of all control rods inserted into the core, by axial
statistical weighting using the axial profiles from the
axial detector strings. The result is compared with the
operative reactivity margin of 1% and an indication is
provided to the operator of any shortfall so that corrective
action can be taken.The same result is used to assess the
initial reactivity worth available for reactor trip, for
comparison with the required minimum of 0.55% per second.

69. The role of the operator in high power operation is
therefore to:
i.   adjust coolant flow through individual fuel assemblies.
ii.   control axial instabilities by movement of two types of
      absorber.
iii. manually trim the radial–azimuthal power distribution
      to optimise power output, consistent with the automatic
      control by the local automatic rods.
iv.   operate such that the operative reactivity margin is
      maintained and therefore the void and power coefficient
      are within the safety design limits.

v.   in conjunction with (iv), probably maintain effective controller action by holding the automatic control rods within prescribed insertion limits.

vi   in conjunction with (iv), maintain the initial trip requirement of at least 0.55% reactivity worth per second.

70. The local automatic control does not function below 10% power, because of uncertainties in the analysis of the in-core detector signals. The automatic mean power regulating group functions in the high power form, but switches to a group of 4 rods responding to signals from detectors moved into the side reflector at powers below 6%. Note that the uncertainty in the in-core detector readings at low power must also apply to the axial profile used in the assessment of the operative reactivity margin and the initial trip worth, supporting the belief that there is a lower power limit to the normal operating regime.

71. The RBMK design specifically attempts to minimise plant outage, for this reason, there are 5 levels of reactor protection.

Protection level	Action
5	Trip all rods (except the short bottom entry type), proceed to shutdown.
5*	Trip all rods (except the short bottom entry type), until the trip signal has cleared.
3	Rapid ramp down to 20% power
2	Rapid ramp down to 50% power
1	Rapid ramp down to 60% power

72. Note that the trip action motors the rods into the core at 0.4 metres/second, giving a time to rod bottom of the order of 18 seconds from any fully withdrawn rods.

73. The protection level 5 is activated by:
A power overshoot of 10% of nominal full power
A reduction in the reactor period to 10 seconds
Drum separator level high or low.
Low feedwater flow
Excess pressure in leaktight compartments; drum separators, reactor cavity or lower water lines.
Low control rod coolant reserve or flow
Trip of two turbogenerators
Trip of the only operating turbogenerator
Trip of 3 of 4 operating main circulating pumps in either pump room.
Voltage loss in the plant auxiliary power supply system

Failure to respond to protection level 3,2 or 1 demands. Manual trip

74. Protection level 5* is activated by an emergency power overshoot, generating a partial trip, stopping the rods when the overshoot signal clears.The distinction between a power overshoot activating level 5 and an emergency power overshoot activating level 5* is not clear. With such heavy reliance on the operator to maintain a stable configuration and to carry out so many manual rod adjustments, it is believed that level 5* is the normally activated level in response to power overshoot, with full trip by level 5 as the more extreme backup.

75. The protection level 3 is activated by:
Load rejection by both turbogenerators
Load rejection by the only operating turbogenerator

76. The protection level 2 is activated by:
Outage of one of two turbogenerators
Load rejection by one of two turbogenerators

77. The protection level 1 is activated by:
Loss of 1 of 3 operating main circulation pumps in either pump room.
Reduction of water flow in the primary circuit
Reduction of feedwater flow
Reduction of level in the drum separators
Actuation of the group closure key for the coolant circuit throttle regulating valves.

78. The significant differences between the RBMK reactor operation and protection systems and UK practice may be summarised as:

(i) reliance on the operator to maintain the reactor characteristics within the bounds adopted for safety analysis, instead of building in a limited range at the design stage.
(ii) a generally heavy operator work load.
(iii)potential for partial trip and subsequent recovery.
(iv) the long insertion period for rods after trip, instead of dropping the rods under gravity.
(v) as a result of (iii) and (iv), reliance on the operator to maintain an initial reactivity worth for reactor trip of at least β per second.
(vi) there is no alternative shutdown system.
(vii)there is no evidence of any provision of diverse guardlines in the protection system.

Chernobyl — the accident sequence

J. D. YOUNG, BSc, PhD, *Central Electricity Generating Board*

SYNOPSIS. The accident sequence described is based on the Russian report to the Chernobyl review meeting in Vienna, supplemented by discussions with USSR experts after the Conference. It draws heavily on the appropriate section of a report prepared for the IAEA by their nominated experts. It is a factual statement of events other than over the period of the power surge when reliance had to be placed on a mathematical model.

BACKGROUND TO THE ACCIDENT

1. Chernobyl Unit 4 had operated very successfully for three years, with more than 100 reactor-years of operation of this reactor type. Chernobyl was in fact the most successful RBMK unit with an 83% station capacity factor for 1985.

2. The accident occurred during tests designed to determine the extent to which the turbine generator could support essential systems during a major blackout. The intention was to cut off the steam supplies to one of the turbogenerators and use the inertial rundown to supply energy to the reactor's fast acting emergency core cooling system (ECCS).

3. Similar tests had already been carried out on the plant when it was found that the voltage on the generator busbars fell off more rapidly than the loss of mechanical energy. The rescheduled tests were trying to eliminate this problem via modifications to the generator's magnetic field regulator. The main coolant pumps were used to simulate the load on the generator.

4. The initiative for the test and the provision of the procedures appears to have lain with electro-technical rather than nuclear experts. The presumption that this was an electro-technical test with no effect on reactor safety seems to have minimised the attention given to it in safety terms. It was stated in the Russian report on the accident that the procedures were poorly prepared in respect of safety, and that authority to proceed by the station staff was given without the necessary formal approval by the station safety technology group.

5.   A series of safety and test procedural violations occurred which resulted in the reactor being placed in a most vulnerable state with regard to the inherent physics characteristics and all but one trip protection blocked, and this was
essentially rendered ineffective by the action taken. If a
single action were to be selected from a long catalogue of
mistakes as the major contributor to the accident, it must be
the blocking of the reactor-turbogenerator trip prior to the
start of the test. The decision to do this was based on the
desire to keep the reactor on-load so that a number of voltage
regulators could be tested. The reactor was due for planned
maintenance and if the tests failed there would be a year's
delay before they could be repeated.

6.   The sequence described is based on the Russian report to
the Chernobyl review meeting in Vienna, supplemented by discussions with USSR experts after the Conference. This paper draws
heavily on a report prepared by nominated IAEA experts using
the above information*.

THE ACCIDENT SEQUENCE

The planned sequence of events

7.   No details are available of the test procedures of the
earlier measurements.

8.   If it is assumed that similar plant conditions were
being set up for both sets of measurements, then the common
features would be: the slow power reduction to 50% (12 hours);
switching off one turbogenerator; the main coolant pump configuration; blocking the ECCS; reducing power to 20 to 30%
prior to tripping the remaining turbogenerator. The differences would be the need to balance the reactor at this power
level and veto the reactor/turbogenerator trip in order to
hold this power level to enable a number of voltage regulators
to be tested.

9.   Protection was dependent upon an overpower trip which
would have been set at 10% above the balanced power level.

The actual sequence - for full details see table 1.

10. At 01:00 on 25 April preparations for the tests started,
(a)†. At 13:05 the reactor power reached 50% and turbogenerator No. 7 was shutdown[b]. Shortly after, (14:00), the ECCS
was isolated[c]. The power reduction was stopped on request
from system control[d] and not resumed until approximately
nine hours later - figure 1.

11. During this period the ECCS remained isolated[e].
Although subsequent events were not greatly affected by the
delay, the fact that the ECCS was not re-poised reflects the
attitude of the operating staff in respect of violation of
normal procedures.

*I am grateful to my colleagues in this exercise,
Dr. B. Edmondson (CEGB) and Drs. A. Brown and G. Frescura
(Ontario Hydro) for allowing me to use this material.

†Superscripts refer to the sequence in table 1.

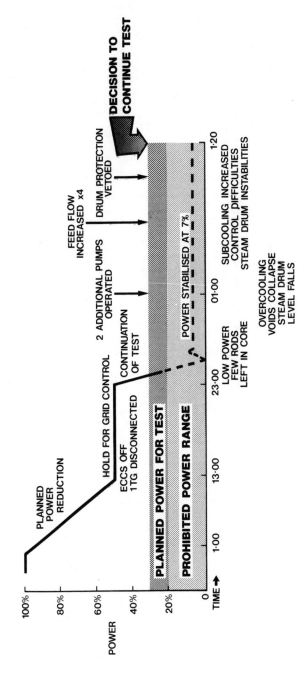

Figure 1. Pre-test Power History

Table 1. A[...] Event Sequence

Time (1)	Event (2)	Interpretation	
		Result of Event (3)	Significance (4)
25 April 1.00.00	The start of reactor power reduction.	Initial steps of test programme and planned maintenance outage.	The slow reduction in power would help to reduce the Xenon poison buildup effects.
13.05.00	Reactor power reduction stopped (b) at 50% full power. Turbine generator No. 7 switched off. Electric power requirements for unit's need switched to TG No.8 (four main circulating pumps, two feed pumps, plus other equipment).		These components will run down with the T.G. during the test. Pump configuration at this time: 4 Running from T.G. 8 2 Running from grid 2 on standby connected to grid
14.00.00	The ECCS was isolated (c).	Done in accordance with the test plan because they wished to avoid spurious triggering.	Safety principle violation but blocked ECCS played no role in transient to point of core disruption. May have been useful post-disruption period.

(1)	(2)	(3)	(4)
1.03.00	The fourth main cooling pump (i) powered from the grid was connected to the left-loop of the heat transport system.	Due to low power and increased flow rate of coolant in the heat transport system the coolant temperature approached saturation.	This adds negative reactivity to the system necessitating more rod withdrawal to compensate.
1.07.00	The fourth main cooling pump (j) powered from the grip was connected to the right loop of the heat transport system.	The flow rate in some of the main cooling pumps exceeded the permissible value. There were significant deviations of the water level and steam pressure in the steam-drums.	Violation of flow vibration limits based on potential cavitation problems. Addition of both pumps caused further control rod withdrawal and further decrease in reactivity reserve margin.
1.19.00	Operator increased feedwater (k) flow. About this time operator blocked the shutdown signals associated with steam-drum level and pressure.	Core subcooling increase results in more void collapse. Control difficulties throughout this period.	
1.19.30	Onset of steam drum water level rise. The feedwater flow exceeded three times balanced value and core void. The automatic control rods went (l) up to the upper tie plate. The manual control rods were (m) pulled up. The onset of SD steam pressure drop.	The feedwater entering the system exceeds the steaming rate. The cooler water reached the core and decreased the steam quality.	Steam drum level increases. Calculated core average void is now zero. Addition of negative reactivity compensated by rod withdrawal.

(1)	(2)	(3)	(4)
14.00.00 (cont.)	Load dispatcher halted the (d) power reduction. ECCS remained isolated (e).		Note: Discussion by USSR experts during the meeting confirmed that ECCS blocking was not necessary for the test. The long hold in power would further reduce the Xe buildup rate at test power level.
23.10.00	Power decrease continued (f) towards the target level of 700-1000 MW(th).		In accordance with test procedure. This level was chosen to be above the minimum allowable operating power of the reactor ($\sim$ 700 MW(th)).
26 April 0.28.00	Operator error with transfer (g) from local (LAR) to global power (AR) control - hold power at required level not entered.	Power reduced to 30 MW(th) due to inability of the automatic control rods and lack of prompt operator action to compensate the void due to power flow mismatch.	Negative reactivity added to system and more manual rods withdrawn to compensate.
1.00.00	The reactor was stabilized at (h) a power of 200 MW(th).	Reactor operating below the minimum allowable power level. Required reactivity reserve margin violated.	No excess reactivity available to raise power.

(1)	(2)	(3)	(4)
1.19.58	The steam bypass valve was (n) closed.	Slow down in the rate of the steam pressure drop.	
1.21.50	The feedwater flow exceeded four times the balanced flow rate.	Steam drum level still rising, pressure still falling.	Control rod position constant as modelled. Reduction in pressure produces enough void to compensate additional feed-water flow.
	Operator abruptly decreases (o) the feedwater flow.		
1.22.10	Steam quality starts rising, (p) automatic control rods start driving in, steam drum water level stabilizes.	Warmer water reaching core inlet produces a rise in average core void, control rods drive in to compensate.	
1.22.30	Feedwater flow reduced to two-thirds of the balanced flow rate.	Operator unable to stop feed-water flow rate at desired level due to coarseness of control system not designed for this operating regime.	Control rods have moved in to compensate added reactivity of increased voiding.
	The distribution of power (q) density and the positions of every control rod were printed out.	This was done to establish the flux distribution and reactivity margin prior to beginning the test.	Confirmation that the opera-tional reactivity reserve margin was half the minimum permissible one, and the operator should have initiated immediate shut-down based on computer printout.

(1)	(2)	(3)	(4)
1.22.45	Feedwater flowrate stabilized.	Steam quality in the core is stabilizing, pressure starts rising.	
1.23.04	The personnel blocked the (r) two-TGs-trip signal. Emergency stop valve to the (s) turbine was closed. The reactor continues operating at a power of 200 MW.	TG No. 8 test starts.	Removal of last process safety system trip to allow test to be repeated. This trip would have saved the reactor. Operator aware he was inducing transient which required shut-down. (This was not provided in the test programme.)
1.23.10	One group of automatic control rods starts driving out.	Core void decreasing because of increasing system pressure.	
1.23.21	Two groups of automatic control rods begin re-insertion.	Reduction of the coolant flow rate and the approach of the warmer water to the core.	Both of these results lead to positive reactivity addition to the core. Control rods trying to balance this addition.
1.23.31	Net reactivity increasing (t) with subsequent slow increase in reactor power.	Control rods can no longer balance added reactivity.	Power slowly rises, positive power coefficient accelerates reactivity imbalance.
1.23.40	Operator pushes AZ-5 button. (reactor trip)	No apparent effect.	

12. At 23:00 power reduction resumed[f] and another unusual event took place. As the operator transferred unit power control from the local to the global regulator system a "hold power" request was not entered. Consequently the reactor power decreased rapidly[g] below the minimum permissible level of 700 MW(th) as a consequence of the collapse in voids (reduced boiling) in the reactor. The reactor power fell to 30 MW(th) and the operator was only able to bring it back to 200 MW(th) by manually withdrawing some of the control rods.

13. In the power regime below about 700 MW(th) the relationship between steam volume and mass is highly non-linear, to the point where small power (and hence steam mass) change leads to large steam volume (and hence void) changes making power control and feedwater control very difficult. If, in addition, too many control rods are withdrawn from the core to maintain balance, this creates conditions which both accelerates the reactor's response characteristics to plant or reactor perturbations and reduces the effectiveness of the protection system.

14. The position of control rods are of dominant importance in determining these response characteristics. The further they are withdrawn, e.g., to maintain a constant power, the more positive the void coefficient and the more sensitive the reactor to any effect that results in a change in the void distribution and/or level in the core. If the power changes, the fuel heating up and the conduction of the heat into the coolant modifies the rate of rise. The fuel temperature coefficient is negative and the net effect of the fuel temperature rise and the additional voiding resulting from that rise is dependent upon the power level. For the RBMK reactor under normal conditions, the net effect (power coefficient) is negative at full power and becomes positive below about 20% power. If operational manoeuvres make the void coefficient larger than normal then this has a direct effect on the magnitude of the power coefficient and the power range over which it remains positive.

15. With the reactor now operating at a level which was not normally allowed, the operator decided to continue the test programme. While this was a major violation of operating procedure, it was not in itself enough to have caused the accident. Later errors compounded the problem.

16. The operator turned on the fourth pump in each loop, according to plan[i] but because the reactor was in such a low void condition with low system thermal hydraulic resistance, the pumps delivered excessive flow to the point where they exceeded their allowed limits derived from cavitation considerations. The throttling valves could not be trimmed further and the increased core flow led to steam drum level problems. The operator compensated by increasing feedwater flow but was unable to get the steam drum level to the desired value because of the coarseness of feedwater control at these power levels.

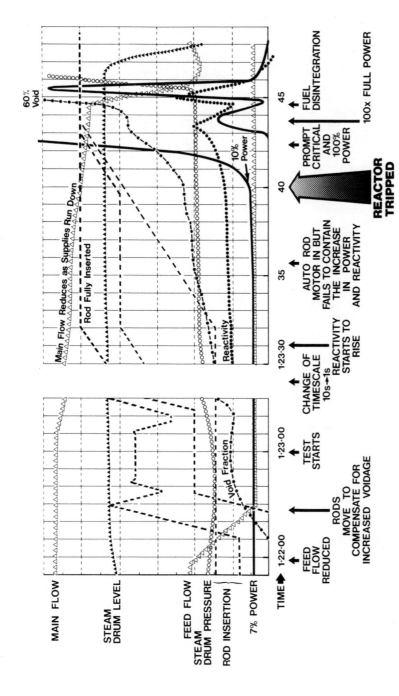

Figure 2. Fault Transient

17. By 01:19 on 26th April the steam drum level was still hovering near the emergency level. The operator increased feedwater flow[k]. Introduction of this feedwater, while raising the drum level, resulted in void reduction which added negative reactivity to the system. The automatic control rods attempted to compensate[l] but required further withdrawal of manual rods to maintain the reactivity balance[n]. The system pressure started falling and steam to the turbine bypass was shut off in an attempt to stabilise pressure[r]. The relationship between pressure, water flow and voidage are all interlinked so that they determine the control system actions. The system response to these interactions is further sensitised by the low power condition. Since the operators were having trouble with pressure and level control, they switched off the reactor trips associated with these parameters[k].

18. It should be noted that from around this point onward (01:19, 26 April) much of the information presented is based on calculation provided by the USSR delegates.

19. When the operator decided the steam drum level was sufficiently high he sharply reduced the feedwater flow[o], producing voidage and more positive reactivity, and the control rods were inserted automatically to compensate and maintain reactor power constant[p]. The operator obtained a printout of the flux distribution from the station monitoring computer before the test. This printout showed that too many control rods were out of the core and that he did not have enough reactivity reserve to meet his shutdown requirement[q] - figure 2.

20. At this time he should have shut down the reactor. The reactor trip on loss of the second generator was switched off to allow repeat of the test if needed[r]. This is a key violation of the test programme, since the reactor would have safely shutdown when the test began, even with the rod configuration existing at the time. The test could, and should, have been conducted in such a way that the reactor tripped when the test began. Such a procedure had worked successfully on previous tests. The only reactor protection signals left in operation were the trips of high power and low period.

21. When the emergency stop valve to the turbine was closed (s), the steam pressure began to rise, the flow through the core started to drop because four of the main cooling pumps were running down with the generator. The increasing pressure, reduced feedwater flow and reducing flow through the reactor are competing factors which determine the volumetric steam quality and hence power of the reactor. It should be emphasised that the state the reactor was now in was such that small changes in power would lead to much larger changes in steam void, and hence power increases. The combination of the factors ultimately led to a power increase beginning at about 1:23:30[t].

22. The shift foreman ordered shutdown of the reactor at 1:23:40 just before the reactor's overpower trip was triggered, but by that time it was too late. There was insufficient

reactivity left in the rods which were in core and the others at the top of the core could not be inserted fast enough to overcome the power increase caused by the competing factors cited above. The positive void coefficient of reactivity inherent in the RBMK design, coupled with the positive power coefficient, continued to add more reactivity and the prompt critical value was exceeded. Within four seconds after 1:23:40 the power was calculated to be one-hundred times full power. This catastrophic increase in reactor power resulted in fuel fragmentation, rapid steam generation and ultimate destruction of the reactor core and associated structures.

EXTENT OF THE DAMAGE

23. The extent of the damage is based on model predictions, visual observations and post-accident on-site measurements.

   (i) It is assessed that approximately 30% of the fuel fragmented with fuel temperatures reaching 4000 to 5000°K. Fuel was ejected from the core. The major part of the fuel seems to be below the reactor space, with a portion above the core and in adjacent rooms.

  (ii) It is assessed that approximately 25% of the graphite blocks were ejected from the core with some blocks outside the reactor hall.

 (iii) The 1000-ton upper plate lifted shearing off the pipes attached to the primary circuit. It is now standing upright.

  (iv) The graphite fire that followed is estimated to have consumed some 10% of the graphite.

   (v) The Russian experts believe that the chain reaction was terminated through fuel homogenisation and relocation.

THE PROPOSED RBMK DESIGN CHANGES

24. Explanation of the fault sequence has been based on an analytical model tested against the plant data available before the accident. The proposed design changes are based on this understanding and it is important that these are both correct and sufficient. Sufficiency must be judged not only against the severe fault context but against more common fuel failure modes that have the potential to release significant quantities of activity into the circuit and to the environment. These transients are orders of magnitude lower than the one that destroyed Chernobyl. The proposed plant modifications will alter the frequency/risk assessment in a fundamental way which could take the RBMK designers some time to re-evaluate.

25. If examination of the nature and extent of the structural and fuel element damage can place 'bounds' on the range of the power excursion, then the analytical models used are asked to bridge the gap between the known pre-accident state and these bounds. These constraints have been used, and the

Russian analysis suggests that at this stage it is possible to explain the damage in terms of the reduction in main flow and the void and power coefficients set up by the test conditions. Any underestimate of the flow reduction (e.g., if pump cavitation had occurred) would need to be associated with a systematic decrease in the void coefficient - if the same reactivity driving function is to be conserved. Whether it should be, depends upon the uncertainties on the bounds of the power rise and thermal-to-mechanical energy conversion processes.

26. The analysis of the transient is very complex and the rate of build-up of power is sensitive to many factors. The analytical models used are still being developed and the conclusions could alter. For example, if the power surge were localised (which, though not ruled out, was declared to be unlikely) then the effective void and power coefficients in that local region would need to be more positive than current calculations suggest - or the flow/pressure reduction greater than currently assumed. Resolution of these and similar issues would reduce the margin of uncertainty in quantifying the effect of the proposed plant modifications.

27. It is not possible to judge whether the modifications are sufficient without having access to a large amount of proprietary information and a large computational exercise. The proposed changes are all desirable and move the inherent characteristics of the plant in a direction which makes its design significantly more robust to operator action and component failure.

28. All control rods will have a minimum insertion of 1.2 m through limit switches. This will be combined with an increase in the reactivity reserve margin to 70-80 rods with automatic trip if this margin is removed. Not only will these increase the shutdown rate but through making the void coefficient more negative, reduce the demand on that rate. In addition, power operation below 700 MW(th) will be prohibited via trip protection.

29. In the longer term, reduction in the magnitude of the void coefficient will be achieved through an increase in the fuel enrichment to 2.4% and the use of fixed absorbers to take up the excess reactivity. Both these measures are robust engineering solutions to the void coefficient problem.

30. It was announced during the Conference that a fast shutdown system was being developed. This is most welcome and it is hoped that the system can be engineered to the required capacity and reliability.

31. It is claimed that these modifications would ensure that a positive overshoot of reactivity for any change in reactor coolant would not cause a prompt critical excursion. This is not the case in the current design and would ease the demands on their protection systems.

REFERENCE
USSR State Committee on the Utilization of Atomic Energy.
The Accident at the Chernobyl's Nuclear Power Plant and its
Consequences.
Information compiled for the IAEA Experts' Meeting,
25-29 August 1986, Vienna.

ACKNOWLEDGEMENT
This paper is published with the permission of the Central
Electricity Generating Board.

The Chernobyl accident — source terms and related characteristics

P. N. CLOUGH, MA, PhD, *Safety and Reliability Directorate, UKAEA*

SYNOPSIS. The Chernobyl reactor accident gave rise to a large release of radioactivity to the environment. The detailed characteristics of this release in terms of radionuclide composition, timing, and energy of release, that is the source term, are discussed in this paper. It is important to be able to establish this source term as precisely as possible in order to relate it to the consequences of the accident. The evidence presented by the Russians at the Vienna meeting, 25-29 August, is reviewed, and discussed in relation to evidence from Western European sources. Important phenomena in the progression of the accident which influenced the source term are examined.

INTRODUCTION

1. When, following the first indication of abnormally high levels of airborne radioactivity in Scandinavia on the 28 April 1986, it was confirmed that a major nuclear reactor accident had occurred at Chernobyl in the USSR, a key question was 'How bad is it?'. It could be quickly established that core of an RBMK reactor of the type at Chernobyl, Unit 4, would contain several thousand million curies of radioactivity ('core inventory'). The severity of the accident consequences was clearly closely related to the fraction of this escaping, to the environment. The information which we now have on the detailed activity released and associated characteristics important for determining the consequences of the Chernobyl accident, that is, the source term, is the subject of this paper. The definition of the source term involves three main ingredients:-

(a) the quantities of specific radionuclides released to the environment. Some 54 individual radionuclides are usually considered to be important for consequence assessment, of which the most prominent are isotopes of iodine, caesium and tellurium. The released activities of these are conventionally expressed as fractions of their inventories in the core at reactor shutdown ('release fractions').

(b) The timing of release, which includes both the start and duration of release relative to shutdown. A warning time for countermeasures may also be required for consequence modelling.

(c) the energy and height of release, which related to the rise and dispersion of the radioactive plume transported from the reactor site.

2.   This definition of the source terms was largely established in the formative study of the risks due to severe accidents in Light Water Reactors (LWR) in the USA, the Reactor Safety Study (1). In this study it was found possible to represent the wide range of severity of potential activity releases in terms of a number of so-called release categories, each with a characteristic source term. Two such source terms for the Surry 1 Pressurised Water Reactor (PWR) are quoted in Table 1. The important radionuclides are grouped into seven groups, the members of each being similar in chemical and physical characteristics, and thus exhibiting similar release behaviour. Each group is represented by a leading element. Thus, Cs covers all relevant isotopes of caesium and rubidium, whilst La covers isotopes of some 8 lanthanide and 4 actinide elements, including plutonium. For each group, the corresponding release fractions in the source terms are shown. Clearly, the most severe source term PWR1 has the highest release fractions, which for the more volatile nuclides (Xe, I, Cs, Te groups) correspond to large proportions of the core inventory. Such a source term will, unless the reaction is sited very remotely, result in very serious consequences. For a PWR such as Surry 1, the containment building must fail almost immediately, or be by-passed in some way, for such a source term to arise. If the containment holds out even for a few hours, a significantly less severe source term results. Indeed, if the containment does not rupture above ground at all, but only suffers basemat failure, a much reduced source terms is expected, represented by PWR6. Such a source term will generally result in much less serious environmental consequences. For comparison, the best estimate which the Russians have been able to provide of the release fractions in the Chernobyl accident up to the 6 May when activity release was finally terminated are also quoted in Table 1. This shows that the Chernobyl accident lies towards the upper end of the spectrum of source terms severity. However, it must be pointed out that the protracted release at Chernobyl (11 days) differs markedly from the much shorter PWR release periods.

THE RUSSIAN ACCOUNT OF THE SOURCE TERM

3.   The Russians have provided a relatively detailed account of their own reconstruction of the source term in the official documents provided to the IAEA Experts Meeting on the Chernobyl accident, Vienna, 25-29 August 1986. The main features of this will be considered first, and then compared with our own attempts to reconstruct the source term on the

Table 1. Source Term Estimates for the Surry 1 PWR compared with Chernobyl

	Start	Duration	Fraction of Core Inventory Released						
			Xe	I	Cs	Te	Ba	Ru	La
Surry 1									
PWR 1	2.5hr	0.5hr	0.9	0.7	0.4	0.4	0.05	0.4	$3(-3)$*
PWR 6	12hr	10hr	0.3	$8(-4)$	$8(-4)$	$1(-3)$	$9(-5)$	$7(-5)$	$1(-5)$
Chernobyl**	0hr	11d	0.8	0.2	0.1	0.15	0.04	0.05	0.02

* $3(-3) \equiv 3 \times 10^{-3}$

**   quoted uncertainty in Russian sources ± 50%

basis of much more remotely collected and scant information. The Russian account contains two main components. Firstly, there is a broad history of the total activity release from the plant which is related to the variations in the known or assessed state of the core throughout the release period. Secondly, a detailed breakdown of activity releases by specific isotopes at certain stages of the accident is provided, supplemented by various isotope-specific sampling measurements at different locations and times. The latter information is chiefly of interest as exemplifying the type of data the Russians have employed in their source term reconstruction.

4.   According to the Russian account, the activity release can be roughly divided into four stages. The day-by-day releases of activity which they estimate are shown in Fig 1. It is important to note that these values are all adjusted for radioactivity decay to day 10 of the accident (6 May). Thus, the actual release on day 0 (26 April) is quoted as 20-22MCi

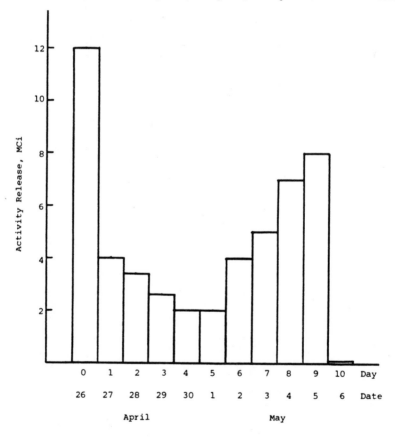

Fig. 1. Soviet account of day by day activity release, corrected to 6 May

by the Russians, but by day 10 only 12 MCi of this remained in the environment following decay. The stages of release were:-

5. Stage 1 (day 0). This was the initiating in-core explosion at 01:24 on the 26 April. Significant quantities of fuel were ejected in the transient which blew off the pile cap, and these were accompanied by enhanced releases of the volatile fission products iodine, caesium and tellurium. The estimated isotope-specific activity releases on the 26 April are shown in the first column of Table 2 (2), but it is not clear what the contribution of the initial explosion was compared with the fire which followed.

6. Stage 2. (Days 0-6). Following the initial explosion, the core top was left fully exposed to the atmosphere, and it appears that on intense graphite fire rapidly developed. High levels of activity release were associated with this during the first day, attributed by the Russians at least in part to the re-entrainment of fuel material embedded in the graphite

Table 2. Soviet assessment of the radioisotopic composition of the release from Chernobyl Unit 4 (ref 2)

Isotope	Activity of release, MCi		Fraction of activity released from the core by 06.05.86,* %
	26.04.86	06.05.86	
Xe133	5	45	Possibly up to 100
Kr85m	0.15	–	"
Kr85	–	0.9	"
I131	4.5	7.3	20
Te132	4	1.3	15
Cs134	0.15	0.5	10
Cs137	0.3	1.0	13
Mo99	0.45	3.0	2.3
Zr95	0.45	3.8	3.2
Ru103	0.6	3.2	2.9
Ru106	0.2	1.6	2.9
Ba140	0.5	4.3	5.6
Ce141	0.4	2.8	2.3
Ce144	0.45	2.4	2.8
Sr89	0.25	2.2	4.0
Sr90	0.015	0.22	4.0
Pu238	$0.1.10^{-3}$	$0.8.10^{-3}$	3.0
Pu239	$0.1.10^{-3}$	$0.7.10^{-3}$	3.0
Pu240	$0.2.10^{-3}$	1.10^{-3}	3.0
Pu241	0.02	0.14	3.0
Pu242	$0.3.10^{-6}$	2.10^{-6}	3.0
Cm242	$0.3.10^{-2}$	$2.1.10^{-2}$	3.0
Np239	2.7	1.2	3.2

* Evaluation error ±50%.

during the initial excursion. During the 27 April, dumping of materials onto the core debris from the air began to extinguish the fire, and by the 10 May some 5000 te had been dropped. In addition to boron compounds and lead, the materials included dolomite, clay and sand specifically intended to trap and filter out the active species from the core. This approach was reasonably effective, because during the 27 and 28 April the activity release rate was substantially reduced, and remained at a reduced level until the 2 May. However, release during this period was still very significant, and it is unclear whether the in-core fire was fully extinguished.

7. <u>Stage 3</u> (Days 7-9). A marked increase in activity release occurred on the 3 May, which continued through the 4 and 5 May. This is attributed to the decay heat pushing the temperature of the core debris (fuel and moderator) to elevated levels, ultimately in excess of 2000°C, following, the effective insulation of the pile up by dumped material. During the early part of this stage, strong enhancement of the volatile fission product contribution to activity release, notably of iodine, was observed. This is clear from the data reproduced in Table 3 (3), which shows the isotope-specific activity distribution in air samples from above the reactor at various dates. However, by the 5 May when the temperature peaked, the released activity assumed a composition close to that of the fuel itself. There is speculation in the Russian account that reduction of the UO_2 by the graphite moderator at these elevated temperatures may have contributed to the fuel-like release. Continued graphite combustion is also referred to, which is incompatible with fuel carbidisation.

8. <u>Stage 4</u> (Day 10). On the 6 May, the release rate rapidly fell to effectively zero. This is attributed to a combination of factors which somehow brought about a rapid cooling of the fuel debris. A key feature seems to have been the injection of nitrogen at a high rate underneath the core, which simultaneously cooled the debris and stifled any residual graphite fires. Additionally, there is reference to special measures taken to promote the formation of more refractory fission product compounds by the introduction of further materials into the filter bed, but this is not enlarged upon. The precise reason why activity release fell so rapidly on the 6 May remains something of a mystery.

9. The total activities contributed by specific isotopes released from Chernobyl and still present in the environment on the 6 May, according to Russian estimates, are shown in the second column of Table 2. These values allow for decay between reactor shutdown and the 6 May. Thus, although 4 MCi of Te 132 was released on the 26 April, the residue of this plus all subsequently released Te 132 only amounted to 1.3 MCi in the environment on the 6 May. This exemplifies a problem in representing a protracted activity release such as that at Chernobyl by means of release fractions as is conventional in source term methodology. The release fraction approach was

Table 3. Relative activity of radionuclides in the air
above the Chernobyl nuclear power plant (ref 3)

Nuclides	April 26	April 29	May 2	May 3	May 4	May 5	Relative Shutdown Inventory
^{95}Zr	4,4	6,3	9,3	0,6	7,0	20	3,6
^{95}NB	0,6	0,8	9,0	1,3	8,2	18	3,8
^{99}Mo	3,7	2,6	2,0	4,4	2,8	3,7	3,9
^{103}Ru	2,1	3,0	4,1	7,2	6,9	14	3,9
^{106}Ru	0,8	1,2	1,1	3,1	1,3	9,6	2,1
^{131}I	5,6	6,4	5,7	25	8,2	19	2,3
^{132}Te + ^{132}I	40	31	17	45	15	8,6	6,4
^{134}Cs	0,4	0,6	0,6	1,6	0,6	-	0,6
^{136}Cs	0,3	0,4	0,5	0,9	-	-	0,I
^{137}Cs	-	-	1,4	3,7	1,3	2,2	0,4
^{140}Ba	3,2	4,1	8,0	3,3	13	12	3,8
^{140}La	11	4,7	15	2,3	19	17	4,0
^{141}Ce	1,4	1,9	7,6	0,9	6,4	15	3,6
^{144}Ce	1,6	2,4	6,1	-	5,1	11	3,4
^{147}Nd	1,4	1,7	2,5	-	2,1	5,4	1,4
^{239}Np	23	3,0	11	0,6	2,8	6,8	56,7

developed in the context of releases of short duration
occurring at most a few hours after reactor shutdown. One
approach for Chernobyl is to quote cumulative release
fractions at the end of release on the 6 May. These would be
given for each isotope by dividing the environmental burden at
that date by the calculated inventory in the intact core at
that date, allowing for decay in the core since shutdown.
Such values calculated by the Russians are quoted in the third
column of Table 2, and utilised in Table 1. However, these
fractions are of limited value for consequence assessment,
especially as the isotopic composition of the release varied

substantially with time. What is needed, as a minimum, is the day-by-day actual activity releases due to each radiologically important isotope. In this context, the concept of release fractions loses much of its usefulness.

10. A further problem in establishing release fractions relates to defining the detailed core inventory of the Chernobyl reactor at shutdown. This is also important for assessing the isotopic composition of the many environmental samples which have been measured, and relating these to the mechanisms of release from the core. The Russian documentation provided at Vienna contained no direct core inventory information. However, the final column of Table 3 quotes Russian results for the relative activities contributed by a set of important radionuclides at shutdown. To obtain absolute data, and to provide a cross-check on various aspects of the Russian analysis, the UKAEA core inventory code FISPIN has been run to estimate the Chernobyl inventory. Full-power operation up to 12 hours before the accident was assumed, followed by the power history specified during the experimental period which led up to the accident. The quoted average fuel burn-up of 10,3000 Md/te was employed, with no allowance for burn-up variations across the core. A Magnox-type neutron spectrum was assumed in the absence of details of the Chernobyl neutronics. In view of these simplifications, the results of this FISPIN calculation must be taken as indicative rather than definitive of the Chernobyl inventory. The FISPIN relative activities for the set of nuclides in Table 3 are compared with the Russian values in Table 4. The agreement is not very good even for the major fission products such as I131 and Cs137. Large discrepancies exist for Ru106 and Cs134, although the latter which is formed by neutron capture is rather sensitive to the assumed neutron spectrum.

11. As an indication of whether the FISPIN or Russian data on inventories are more reliable, Table 5 compares the predicted ratios of isotopes of the same element with the measured ratios in airborne samples taken from Table 3. Clearly, the release fractions of isotopes of the same element must be identical, so that the measured ratios reflect the ratios in the core. The FISPIN Cs134/Cs137 ratio is about 40% higher than the average sample value, but the Russian result is a factor of 3.3 too large. For the ruthenium and cerium isotopes, the FISPIN results tend to under-predict the ratios, and the Russian values to over-predict them, although in neither case are the discrepancies really serious in view of the measurement uncertainties. On the whole the FISPIN results give a reasonably good account of the data on isotopic ratios, and are used as the basis for discussion of the sampling measurements below. However, a further anomaly in the Russian reporting of the activity release relating to inventory and decay should be pointed out. The total activity release for the 26 April in the first column of Table 2 amounts to just over 20 MCi. Correction for decay to the 6

Table 4.   Comparison of FISPIN and Soviet Shutdown
           Inventories

	Activity (FISPIN) MCi	% of subtotal	Soviet % of subtotal	Soviet / FISPIN ratio
^{95}Zr	156	5.6	3.6	0.64
^{95}Nb	159	5.7	3.8	0.66
^{99}Mo	148	5.3	3.9	0.74
^{103}Ru	116	4.1	3.9	0.95
^{106}Ru	23	0.84	2.1	2.5
^{131}I	81.1	2.9	2.3	0.79
^{132}Te + ^{132}I	221	7.9	6.4	0.81
^{134}Cs	4.05	0.144	0.6	4.17
^{136}Cs	2.16	0.077	0.1	1.30
^{137}Cs	6.48	0.231	0.4	1.72
^{140}Ba	154	5.5	3.8	0.69
^{140}La	159	5.7	4.0	0.70
^{141}Ce	148	5.3	3.6	0.68
^{144}Ce	105	3.7	3.4	0.92
^{147}Nd	59.5	2.1	1.4	0.67
^{239}Np	1319	45.0	56.7	1.26
Subtotal	2866			

Table 5.   Comparison of calculated and measured isotope
           ratios in Chernobyl release

	Calculation		airborne above reactor			
	FISPIN*	Soviet	day 0	day 6	day 7	day 8
Cs $\frac{134}{137}$	0.63 (0.63)	1.49	–	0.43	0.43	0.46
Ru $\frac{106}{103}$	0.21 (0.24)	0.54	0.38	0.27	0.43	0.19
Ce $\frac{144}{141}$	0.70 (0.84)	0.94	1.14	0.80	–	0.80

* value at shutdown;   day 10 value in brackets.

May by the Russians results in the 12 MCi value which appears Fig 1 for day 0. An independent decay calculation for the set of nuclides quoted in Table 2 gives a decay factor of 3. Thus, the Russian estimate for release on the 26 April of 20 MCi compares with 36 MCi back-calculated from Fig 1.

RADIONUCLIDE SAMPLING DATA FROM WESTERN EUROPE AND THE USSR

12. Radioactivity sampling data from Scandinavia was supplied to SRD soon after the enhanced levels in Finland and Sweden were detected following the Chernobyl accident and an exercise to establish the source term began. Subsequently, widespread sampling data from around Western Europe came in. The absolute levels of activity measured were important for estimating the magnitude of the source term. Based on the Scandinavian data, the atmospheric dispersion and consequences modelling code CRACUK was applied to back-calculate the likely size of the release once the site of the accident had been confirmed. Release fractions of about 5% for the volatile fission products iodine and caesium were deduced for the release towards Scandinavia. These values compare favourably with the 5-10% release fractions for these fission products quoted by the Russians for the 26 April, when the plume was directed towards Scandinavia.

13. Some of the Western European data contained radionuclide specific activity measurements, and these were particularly important for trying to reconstruct the state of fuel and core damage, and thus to establish details lacking at that time on the nature of the Chernobyl accident. The volatile fission products (noble gases, iodine, caesium, tellurium) are released from UO_2 fuel at lower temperatures, typically 1500-2000°C, than the more refractory fission products such as barium, strontium, ruthenium, lanthanum, cerium, and the actinides. The latter require temperatures well in excess of 2000°C to undergo significant release. Moreover, certain fission products are sensitive to the oxidizing conditions around the fuel. For example, tellurium becomes strongly bound to the zirconium in the fuel cladding when first released from the UO_2, but if the cladding itself becomes oxidized, the tellurium is freed. Similarly, ruthenium as an element is highly refractory, but in oxidizing conditions, much more volatile oxides can be formed. Enhanced releases of tellurium and ruthenium would indicate oxidizing conditions in the core, such as would be associated with an in-core fire.

14. Now that the Russian account of the accident is available, it is interesting to re-examine the Western sampling data and to relate it to the phenomenology of the accident as described by them. Radionuclide specific sampling data collected in Finland, Sweden, Denmark, the Netherlands and Hungary over the two weeks following the accident gave a remarkably consistent picture of the radionuclide spectrum in the release. The averaged results of air sampling measurements are shown in the first column of Table 6. The

values quoted are the release fractions for individual
isotopes relative to the release fraction for Cs137. The
FISPIN-calculated core inventories were used to generate these
fractions. Thus,

Relative Release Fraction for X =

$$\frac{\text{Activity of X in sample}}{\text{Activity of X in core}} \bigg/ \frac{\text{Activity of Cs137 in sample}}{\text{Activity in Cs137 in core}}$$

All activities are corrected for decay back to the time of
reactor shutdown. It is evident that the release fractions of
the volatile fission products inferred from these samples are
one to two orders of magnitude higher than those of the more
refractory species.

15. It is now apparent that the activity transported to
Scandinavia during the two days following the start of the
accident must have been associated with the initial peak of
release on the 26 April (Stage 1 plus the first day of Stage
2). A change of wind from north westery to south westery then
transported the residue of the same material towards the Low
Countries. This accounts for the similarity in radionuclide
composition observed in these regions, and indeed later in the
UK. One may compare this composition with that in air samples
above the reactor on the 26 April. (Table 3). The relative
release fractions for a selection of isotopes are shown in the
second column of Table 6, normalised to Cs134 since Cs137 was
not measured. The ratio of volatile to more refractory
nuclides is much reduced relative to the remote samples,
although there is still significant enhancement of the

Table 6.   Comparison of relative release fractions
            from remote and local airborne samples

Element	Isotope	Relative Release Fractions	
		W European Samples	Chernobyl 26.4.86
Cs	137	(1.0	1.0
	134	(	
I	131	(	
	132	(0.35 − 0.50	0.72
	133	(	
Te	132	0.55 − 0.90	1.84
Ba	140	0.035 − 0.050	0.21
La	140	0.005	
Ru	103	0.055 − 0.120	0.19
	106		0.34
Ce	141	0.005	0.09
	144		
Nb	95	0.005	
Np	239	0.002 − 0.014	0.19

volatile content compared with the whole fuel composition. The activity quoted by the Russians for Te132 includes a contribution by I132, which partially accounts for the anomalously high relative release fraction assigned to Te132.

16. The differences in remote and local sample compositions may have arisen during the initial stages of the accident as follows. High fuel temperatures were undoubtedly reached in the initial reactivity transient. Russian estimates of the temperature history of the fuel throughout the accident are shown in Fig 2. The initial transient, which generated sufficient pressure to rupture most of the pressure tubes and severely disrupt the pile up, drove fuel temperatures in the zones of peak reactivity to levels where significant fuel fragmentation occurred. At these temperatures, the volatile fission products were almost entirely released from the fragmented fuel. Moreover, other zones of the fuel which did not get sufficiently hot to fragment nonetheless became hot enough to release large fractions of their volatile fission product contents. In the initial explosive burst, a small amount of fuel was ejected from the core ($<0.5\%$ interpolating from Table 2), a proportion of which was in the aerosol size range. This was accompanied by a larger fraction of the vapours of the volatile fission products (a few percent). In the rapid cooling process on exiting the core, the vapours of the volatile species condensed. A proportion condensed onto the fragmented fuel, most of which was of a particle size such that it settled out fairly rapidly from the atmosphere. However, condensation onto fine fuel particles, possibly accompanied by self-nucleating condensation, produced a fraction of very fine aerosol heavily enriched in the volatile fission products. It was this material which was carried over 1,000 km to Scandinavia and contributed heavily to the samples in Western Europe.

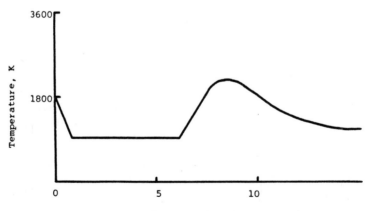

Fig. 2. Soviet estimate of temperature history of Chernobyl fuel

17. In terms of this interpretation, peak fuel temperatures in the hottest zones of the core must have reached or exceed 3000K in the initial transient. In several respects it more closely resembles the initiation of the Hypothetical Core Disruptive Accident in an LMFBR than any type of accident anticipated for Western designs of PWR or BWR. The Russian estimate of 1800K for the fuel temperature at the outset is certainly too low, although this may refer to the average over the core, rather than the hottest region.

18. The Russians have provided no information on the relative contributions of the initial burst and the subsequent in-core fire to the total activity release on the 26 April. The nature of the fire or fires in the core is still unclear, and is discussed in more detail below. From Fig 2, the fuel temperature appears to have fallen quite rapidly to about 1000K over this day, before any effective countermeasures were begun. It is likely that the air sample quoted in Table 3 was taken some hours after the initial explosion. A possible mechanism for continuing activity release in an oxidizing fire regime at relatively low temperatures is the oxidation of UO_2 to U_3O_8. Large amounts of fragmented, unclad, UO_2 pellets must have been present in the core debris after the initiating excursion. Oxidation of UO_2 to U_3O_8 in air promotes activity release by both chemical and physical mechanisms. The chemical mechanism accelerates the release of certain fission products, notably iodine, tellurium and ruthenium from the fuel matrix and is effective over a wide temperature range about 800K (4). The physical mechanism is due to the production of aerosol-size U_3O_8 particulate which is readily spalled away (5). This is effective over the temperature range 800-1000K but at higher temperatures larger particular U_3O_8 is formed which is less mobile.

19. A strongly oxidizing regime in the core at a temperature of around 1000K with ventilation generated by the in-core fires could thus have led to a steady release of U_3O_8 aerosol accompanied by enhanced releases of the volatile fission products. A composition corresponding to whole fuel enriched with volatile fission products is evident both in the air samples from above the reactor (Table 3 and 6), and in ground samples close to the site (Table 7 (6)). The vapours of the volatile species would deposit onto the aerosols rapidly upon cooling after leaving the core. The Russians themselves have attributed this release to re-entrainment of fuel impacted into the graphite as the latter burned, but such an explanation appears unnecessary. During the 26 and early part of the 27 April, the release was unconfined. Thereafter, it became increasingly attenuated as filtering materials were dumped on the core debris. However, these filtering materials were far from 100% effective, since substantial releases persisted throughout Stage 2 of the accident. The filter bed itself probably became very hot, and was thus ineffective for trapping vapours of the volatile species. Moreover, under some conditions the proportion of U_3O_8 particles formed by UO_2

Table 7. Comparison of relative release fractions
from deposition samples in Chernobyl and
Kiev regions

	Relative Release Fractions	
Isotope	Average Chernobyl Soil Samples	Kiev UK Student Shoe*
Cs137	1.0	1.0
134	1.52	0.74
I131	0.89	1.22
Te132	1.00	1.47
Ba140	0.14	–
Zr85	0.12	1.82
Ru103	0.17	1.02
106	0.16	1.38
Ce141	0.13	–
Np239	–	1.02
Pu238	0.010	4.21
Pu(239 + 240)	0.007	3.10
Pu241		1.95
Am241		11.2
Cm242		4.57
Cm244		8.89

* measured by NRPB – private communication

oxidation in the 1μm size range can be significant. Such
small particles would be poorly trapped in a coarse filter.
   20. The chief evidence for strongly oxidizing conditions in
the core comes from the high release fraction of tellurium,
comparable with iodine and caesium. As previously noted, this
indicates extensive oxidation of the zirconium alloy fuel
cladding. High temperature oxidizing conditions would also
lead to the decomposition of caesium iodide, which is thought
to be the predominant chemical form of fission product iodine
in oxide fuel. There is no evidence in any of the Russian
sampling data for a separation of caesium and iodine in the
release, the release fraction for iodine relative to caesium
being close to unity. However, much of the remote air
sampling data showed an iodine relative release fraction well
below unity (Table 5). This sampling was based mainly on
particulate filters, but in one instance where gaseous iodine
was also trapped on charcoal (Nurmijarvi, Finland), some 85%
of the total was detected in this form. It thus appears that
by the time the released iodine reached Western Europe, both
particular and gaseous forms were present, and that the low
apparent relative release fraction in air samples is an
artifact of the sampling method. Infact, ground samples in
Western Europe yield on iodine relative release fraction close
to unity. It is not clear whether both forms of iodine were
released from the site, but the chemistry regime in the core

makes this probable. However, there is no evidence for enhanced ruthenium release relative to the other non-volatile species in the Russian sampling data, and only a slight indication of this effect in the remote sampling data. The oxidative mechanism is not, therefore, fully consistent with the observations.

21. Although the radionuclide composition through Stage 3 of the release remained fairly similar to that through Stage 2 on average, there is evidence that on the 3 May a sharp temporary increase in the relative contributions of the volatile fission products iodine, caesium and tellurium occurred. The Russians attribute the overall increase in activity release during Stage 3 to rising core temperatures (Fig 2), as the decay heat finally made an impact on the massive thermal sink represented by the graphite moderator. The U_3O_8 spalling mechanism discussed above should have become less effective for fuel release as the temperature rose. Increased release rates for the volatile fission products are fully expected if the core heated up to temperatures in excess of 1500°C, particularly since regions of previously unheated fuel probably became involved. The burst in the volatile species on the 3 May may well have been due to re-vaporization of material previously condensed in the filter bed as this also heated up. However, the high level of release of material with composition akin to fuel is difficult to explain. There has been speculation that regions of the debris became hot enough for fuel liquefaction to have occurred. Direct vaporization of fuel and refractory fission products would then be a possible origin of the enhanced releases. However, such a mechanism would be expected to sharply discriminate between species such as barium, for which reducing conditions promote release, and ruthenium for which oxidizing conditions are more effective. There is no evidence for such discrimination in the sampling data. (Table 3).

22. The termination of the release (Stage 4) remains something of a mystery. The most significant feature appears to have been the injection of nitrogen to displace the air in the core. It is possible that graphite combustion was still contributing significantly to pushing up core temperatures at this stage, and that the extinction of this was an important contributory effect. However, the decay heat state represented a major energy source, and an additional effect related to the termination of oxidative mechanisms which were promoting release during Stage 3 seems necessary to account for the abruptness of the ending of release.

23. Much of the foregoing discussions has been based on the available air sampling data, since this provides the greatest detail with respect to the history of the release. The Russians have also provided a good deal of information on ground samples from close to the site. Indeed, the Russian estimates for the absolute magnitude of the release fractions appear to be based largely on extensive ground sampling measurements. Such samples have provided the main source of

information on actinide releases in the accident, although the
Russian data relates to measurements after the termination of
the release, so that a cumulative picture of the radionuclide
deposition at the sampling location is given. There is a
large variability in the radionuclide composition found in
different soil samples, but a representative set of relative
release fractions found for samples within a 30km zone north
of the reactor is given in the first column of Table 7. This
combines data quoted in several tabulations by the Russians
(6). On this evidence, the relative release of plutonium was
substantially lower than that of other refractory oxide
constituents in the fuel, such as cerium and zirconium. This
result is inconsistent with the much higher relative release
of Np239 indicated by the air sampling. It is also at odds
with the limited independent evidence on the nature of the
deposited activity from within the USSR. The second column of
Table 6 shows the composition of the activity on the shoe of a
student who returned from Kiev (80 km from Chernobyl) to the
UK on the 2 May, measured by the NRPB. Here, excluding the
actinides, the composition is very close to that of whole
fuel. The actinides appear to be present in something of an
excess over their expected core inventories, but in view of
the uncertainties in calculating the latter, undue
significance should not be attached to the actual numbers in
the table. The important point is that at least equal release
fractions are indicated for the actinides as for the other
non-volatile radionuclides. The quoted Russian soil sampling
data may be unrepresentative in this respect. The sample
variability has already been noted, and a marked variability
has also been observed in the composition of individual
radioactive particles which have been analysed in the USSR and
elsewhere. For example, particles which were composed almost
entirely of ruthenium were detected in Sweden (7), whilst
particles of almost pure caesium or cerium were found close to
the site. There is no ready explanation for the occurrence of
such particles of almost unique composition, although
formation via selective vapour condensation processes is a
possible origin.

ENERGETICS OF THE RELEASE
24. The initial explosive release of activity from the core
appears to have been a very energetic event. It is outside
the cope of this paper to attempt any assessment of the energy
release associated with Stage 1. A high rate of energy
release must have continued at least into the 27 April, since
the Russians have quoted a plume height in excess of 1200 m on
that day at 5-10 km from the plant. Brief consideration will
be given as to whether chemical energy sources from combustion
could have contributed significantly to the energetics of the
release processes. The yardstick of comparison is the decay
heat in the core debris, which remain very substantial
throughout the whole of the release period. For example, on
the 27 April the decay heat was around 15 MW, and even at the

termination of release on the 6 May it was still at a level of around 7 MW. If efficient mechanisms existed for converting much of this energy into sensible heat of the plume, it would be unnecessary to seek other sources to account for the higher plume rise.

25. Nonetheless, two chemical reactions which may have supplemented the decay heat substantially are the combustion of zirconium, present in large amounts in fuel cladding and the reactor pressure tubes, and of the graphite moderator:-

$$Zr\ (c) + O_2(g) \rightarrow ZrO_2(c) \quad \Delta H°(1000K) = -1091kJ$$

$$C\ (c) + O_2(g) \rightarrow CO_2(g) \quad \Delta H°(1000K) = -395.5kJ$$

The oxidation of zirconium alloy fuel cladding in air undergoes breakaway to a regime of parabolic kinetics at temperatures above 800°C, whilst graphite combustion breaks away at even lower temperatures of 500 - 600°C. However, zirconium oxidation is favoured both thermodynamically and kinetically over graphite oxidation at the temperatures of 1000°C or higher associated with the early stages of the accident. The zirconium was also concentrated in the fuel channels, where ventilation of the disrupted core was presumably best. Some of it was oxidized by steam in the initial transient, generating the hydrogen which was responsible for early ex-core explosions and fires. However, oxidation of the residual metal by air probably played an important part in sustaining high fuel temperatures at the outset. A simple adiabatic calculation suggests that the complete oxidation by air of the zirconium content of a fuel channel could have heated the UO_2 fuel to a temperature of around 4000°C. Thus, even allowing for appreciable heat loss, a strong chemical heating mechanism was available. Moreover, combustion of the zirconium in the core in air at the rate of 10% per day could have provided about 1 MW of sensible heat in the exit gases. It is more likely, though, that the zirconium was consumed very rapidly once air entered the core, contributing strongly to the initial energy spike.

26. Once the zirconium was locally exhausted in regions of the core, graphite combustion provided a further chemical energy source. This reaction releases less than 40% of the energy available for zirconium oxidation per mole of oxygen consumed, and the adiabatic temperature limit for graphite combustion in air is close to 2000°C. Burning of the graphite moderator at a rate of 1% per day would have generated a sensible heat release of 7.6 MW, which is comparable with the decay heat. There is no information available from the Russians as to their estimates of the total loss of graphite in the in-core fires.

CONCLUSIONS

27. The major characteristics of the source term in the Chernobyl accident appear to have been established by the

Russians. Some 10–20% of the core inventory of the volatile fission products iodine, caesium and tellurium had been released from the core over the 11 day period, 26 April to 6 May, together with 3–4% of the refractory fuel constituents, although ± 50% uncertainty bounds are quoted for these values. The Russian account of the source term is consistent with independent evidence from Western sampling measurements in most respects, although some anomalies remain. The Chernobyl source term differed in important respects, especially the length of the release period, from the types of severe accident source terms which have been considered for reactors of Western design. This feature, and the high initial energy release, were important for consequence mitigation.

REFERENCES

1. Reactor Safety Study. An Assessment of Accident Risks in US Commercial Nuclear Power Plants. (Chairman: N Rasmussen). USNRC, WASH-1400. (1975).
2. Soviet Report to IAEA Experts Meeting, Vienna, 25–29 August 1986, Annex 4, Table 4.14.
3. IDEM, Annex 4, Table 4.10.
4. RITZMAN R.L and PARKER G.W. Summary of data on fission product release from UO_2 during oxidation in air. WASH-1400 Appendix VII. Appendix F (1975).
5. IWASAKI M. et al. J Nucl Sci and Technology, 1968, vol 5, 652.
6. Soviet Report to IAEA Experts Meeting, Vienna, 25–29 August 1986. Annex 4. Tables 4.2 to 4.6.
7. DEVELL L. et al. Initital observations of fallout from the reactor accident at Chernobyl. Nature, 1986, vol 321, 192.

Chernobyl — the radiological consequences in the USSR

M. D. HILL, MA, MSc, *National Radiological Protection Board*

SYNOPSIS. This paper summarises the assessment made by Soviet scientists of the radiological consequences of the Chernobyl accident in the USSR, as presented at the Post-Accident Review Meeting in August 1986.

INTRODUCTION

1. This paper describes the radiological consequences of the Chernobyl reactor accident, in terms of early health effects, doses to the population of the USSR, and the counter-measures which were taken during and after the accident to reduce doses to people and radionuclide levels in the environment. The paper is based on information presented by Soviet scientists during the Post-Accident Review Meeting held at the International Atomic Energy Agency (IAEA) in Vienna over the week beginning 25 August 1986. Some of this information was preliminary and is unlikely to be correct in every detail, given the short time in which it was compiled. Nevertheless, it does enable the scale of the consequences of the accident in the USSR to be appreciated.

RADIONUCLIDE TRANSPORT THROUGH THE ENVIRONMENT

2. During the first few days of the accident the wind blew towards the north and north-west, but by 30 April the wind direction had changed, leading to a plume of radioactive material travelling to the south and east. Throughout the period of the release (26 April to 6 May) there was no heavy rainfall in the area immediately around Chernobyl, or in the nearby city of Kiev, because rain clouds moving towards the region were artificially dispersed (presumably by seeding with silver iodide from aircraft).

3. The complex meteorological conditions, the artificial dispersing of rain clouds and the varying characteristics of the release led to a very complex pattern of atmospheric transport and deposition on the ground, both within the USSR and in other countries. In the region of the Chernobyl site a detailed map of radionuclide deposition levels was built up by measuring external dose rates and by analysis of environmental samples (e.g. soil, grass). The pattern of deposition within other regions of the USSR was established by taking

gamma dose rate measurements from aircraft, and by analysing the radionuclide content of soil samples taken at a limited number of locations. This procedure enabled an estimate of the total amounts of radionuclides deposited in the USSR to be made with impressive speed, and this estimate was used by Soviet scientists in deriving the total quantity of radionuclides released. Some other countries have carried out similar exercises and eventually it will be possible to refine the initial estimate of the total release.

4. Deposited radionuclides, particularly iodine-131 and caesium isotopes, entered the terrestrial food chains. The Kiev reservoir, and the rivers draining into it (see Fig. 1) were initially contaminated by direct deposition, and subsequently by run-off. Pine forests close to the site were particularly heavily contaminated because of the large surface area of pine needles.

5. Approximately half of the particulate material released was deposited within a zone of radius 30 km around the reactor site, and about 0.5% of the fuel released was deposited on the site itself. In these areas, resuspension may have led to redistribution of radionuclides, further complicating the observed deposition pattern.

COUNTERMEASURES TAKEN TO REDUCE RADIATION DOSES TO THE PUBLIC

6. On the morning of 26 April, people in the town of Pripyat were instructed to remain indoors, with doors and windows closed. Potassium iodide tablets were distributed by volunteers who handed the tablets directly to individual residents, on a door-to-door basis. As the day progressed, radiation levels in Pripyat began to rise and it became apparent that doses could exceed the lower intervention level for evacuation previously established in the USSR (250 mSv to the whole body), and eventually even the upper intervention level (750 mSv to the whole body), if the population was not moved. Ad hoc evacuation plans were devised; existing ones could not be implemented, partly because they involved transport through areas which were highly contaminated. By early afternoon on 27 April the 45,000 inhabitants of Pripyat had been evacuated, in an operation taking less than 3 hours. Over the next few days the remaining 90,000 members of the public within 30 km of the reactor site were also evacuated, and tens of thousands of cattle were moved to less contaminated areas where they could be given 'clean' fodder.

7. Immediately after the accident, bans were introduced on the consumption of milk, milk products (cheese, cream, butter), leafy vegetables, meat, poultry, eggs and berries in which concentrations of iodine-131 were above derived intervention levels. These levels were based on an intervention level of dose of 300 mSv per year to the thyroid. At a later stage, when caesium-137 and other longer lived radionuclides became dominant, intervention levels for these radionuclides in a wide range of foods were established and enforced. These levels were based on the principle that the effective dose to

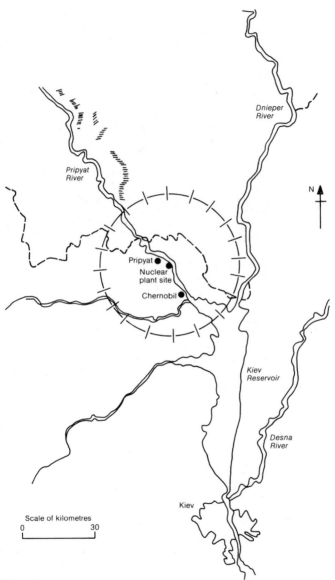

Fig. 1.

an individual should not exceed 50 mSv in the first year after the accident. In the short term, foods with radionuclide concentrations above the derived intervention levels were probably disposed of. However, in the longer term the economic consequences of such actions would be extremely large and thus it is planned that agricultural products will be used for purposes other than direct human consumption, for example feeding animals and as an input to industrial processes.

8. Great emphasis was placed by the Soviet authorities on preventing the consumption of contaminated drinking water and on the protection of the aquatic environment. Radionuclide levels in freshwater, sediments and aquatic biota were carefully and continuously monitored. Breakwaters were constructed in rivers, with the aim of preventing the downstream transport of contaminated suspended particulates. Around Kiev, 400 wells were bored in order to provide new water supplies. The Dneipr River is no longer used as a source of drinking water. At the reactor site, one of the major concerns has been to reduce the probability of the release of radionuclides into groundwater to very low levels.

9. In Kiev, steps were taken to decontaminate buildings and roads. We were told at the IAEA review meeting that, as a result of these actions, Kiev is now "the cleanest town in the world".

10. From measurements of concentrations of actinides (particularly plutonium isotopes) in air in the 30 km zone, it became clear that under dry conditions and if winds were high, doses to individuals from resuspension could exceed the limits normally applied for members of the public in the USSR. Accordingly, measures such as covering of "hot spots", were taken to prevent resuspension. In the longer term, the upper layer of heavily contaminated soil will be removed and disposed of. In less contaminated areas, fixation of radionuclides in soils is planned, the aim being to return the land to agricultural use as soon as possible.

EARLY HEALTH EFFECTS

11. Of the staff on site at the time of the accident and those involved in the immediate on-site response to the emergency, about 300 had to be sent to hospital for treatment. Most of these people were members of the fire brigades who fought the early extensive fires at the site. Acute radiation syndrome of varying clinical severity was diagnosed in 203 people. These symptoms were caused by external gamma and beta irradiation. Internal irradiation due to inhalation of radionuclides did not contribute significantly to these early health effects, and it was established (through measurements of the sodium-24 levels in the body) that there was very little contribution from neutron exposure. Two people died at the reactor site: one was trapped within the building and his body was never recovered, the other died before he could be taken to hospital. No one from the off-site area had to be taken to hospital for treatment of radiation injuries.

12. Table 1 shows the estimated whole body gamma doses to the 203 people who suffered the acute radiation syndrome, and the number of people who had died in the period to the end of July 1986. In most of the cases of early fatality, severe skin burns due to external beta irradiation were a major contributing factor. Twenty people had such extensive burns that between 40% and 100% of their skin was destroyed; these people would have died even if they had suffered no other radiation induced effects, such as bone marrow failure.

Table 1. Whole body gamma doses to the most exposed individuals

DOSE (Gy)	NUMBER OF PEOPLE	
	Exposed	Died
6-16	22	21
4-6	23	7
2-4	53	1
1-2	105	0
Total	203	29

13. The main elements of the treatment of the highly exposed persons were: transfusions of platelets and blood, use of antibiotics and anti-viral drugs, intravenous feeding and establishing antiseptic regimes. Thirteen bone marrow transplants were carried out, but all but one of the patients died. The conclusion drawn by the Soviet doctors was that such transplants are only likely to be effective within a narrow range of radiation doses, and even then the operation can have negative as well as positive effects.

14. In the words of a recent issue of the Lancet, the medical handling of the patients with acute radiation syndrome was, by any standards, first class. Without the specialised centre in Moscow to which most of the patients were taken and which already had experience in treating individual accident cases, it seems likely that many more people would have died in the weeks following the accident.

RADIATION EXPOSURE OF THE PUBLIC

Short term exposure
15. During the period when the plumes of radioactive material were travelling through the atmosphere and radio-nuclides were being deposited on the ground, people living or working in the areas under the plumes received doses via

direct external radiation from the plume, via inhalation of radionuclides and via external irradiation from deposited material. In the most affected areas some doses will have been received through consumption of contaminated milk, leafy vegetables and fruit before measures were introduced to restrict consumption of these products. In areas where the contamination levels were not so high as to require the introduction of bans on food consumption, doses will have been received via all these routes.

16. Initially, estimates of doses were needed in order to take decisions on countermeasures. These were obtained from environmental monitoring data, supplemented by predictive modelling where necessary. At a later stage measurements were made of iodine-131 in the thyroids of people, particularly children, and whole body measurements were carried out to determine levels of caesium-137. These direct measurements enabled better estimates to be made of the doses actually received.

17. The collective dose from external radiation to the 135,000 people who were evacuated was estimated to be $1.6 \ 10^4$ man Sv. Most doses to individuals were less than 250 mSv, although some people in the most contaminated areas may have received doses as high as 300 – 400 mSv or more. Doses to the thyroid of individuals from inhalation (and possibly ingestion of contaminated foods) were estimated to be mostly below 300 mSv, although some children may have received thyroid doses as high as 2,500 mSv. In other regions of the USSR doses to individuals from external irradiation and from intakes of iodine were very much lower.

Longer term exposure

18. Once the release of radionuclides had ceased, and in subsequent months and years, the most important exposure pathways are external irradiation from deposited radionuclides (especially caesium-137) and internal irradiation from ingestion of contaminated food. Iodine-131 is an important contributor to ingestion dose in the first weeks, but in the longer term caesium-137 tends to dominate, although other longer lived radionuclides (such as strontium-90) may also be important. There will also be a contribution to doses from inhalation of radioactive materials which are resuspended into the air from the ground on which they were initially deposited.

19. Many measurements were made of radionuclide levels in the environment, both within the 30 km evacuation zone and in regions in the European part of the USSR (i.e. the Ukraine, Byelorussia and the Russian Soviet Socialist Republic). In the case of the evacuation zone, the corresponding estimates of dose were used to determine whether, and if so when, people could return to the area. For the other regions, the purpose was to determine whether any further countermeasures were required, whether any could be lifted, and to estimate the radiological impact on the population.

20. Doses to individuals in various regions of the European part of the USSR, and doses to the whole population of these regions were calculated by Soviet scientists. Estimated individual doses from external irradiation in 1986 range from 0.03 mSv to 10 mSv. Committed doses from intakes of caesium-137 via foodstuffs were calculated to be of the order 30 mSv for individuals in Byelorussia and the Ukraine. However, measurements of caesium levels in people showed that 50% of them would receive a dose of about 3 mSv or less, and only 3% would receive a dose of 30 mSv or more. The reason for these differences lies in the pessimistic assumptions used in the dose calculations, particularly the assumptions about the transfer of caesium from the soil into plants, and the assumption that the individual consumes only food which is produced locally. Doses tended to be higher for people in rural areas because they spend more time outdoors and eat more locally produced food.

21. The collective dose to the population of the European USSR from external irradiation over the next 50 years was calculated to be about 3.10^5 man Sv. The collective dose to the same population from intakes of caesium in food over 70 years was estimated at 2.10^6 man Sv. During discussion, many experts at the IAEA review meeting questioned this latter figure, because experience in estimating collective doses from releases of caesium to the atmosphere (e.g. from nuclear weapons tests) suggests that the external dose is approximately equal to the dose via food consumption. It was made clear that pessimistic assumptions, of the type mentioned above, had been deliberately used in calculating the collective dose from food. More realistic assumptions could produce an estimate which is a factor of 10 lower.

LATE HEALTH EFFECTS

22. The estimates of the doses to the population of the USSR are preliminary and, to some extent, incomplete; they may well change in the future as a result of more detailed and thorough evaluation. It is therefore not possible at present to make a complete assessment of the number of fatal and non-fatal cancers which may occur in the USSR as a result of the accident. However, a tentative estimate of the number of fatal cancers can be made on the basis of the collective dose estimates presented at the IAEA review meeting.

23. Assuming a no-threshold, linear, dose response relationship, and a risk coefficient of the order of 10^{-2} Sv^{-1}, the number of fatal cancers expected to occur in the 135,000 evacuees as a result of external irradiation would be about 200. The contribution of iodine-131 doses to the thyroid would increase this estimate. For example, if the average dose to the thyroid was about 300 mSv, about 10 cases of fatal thyroid cancer would be expected (plus a substantially larger number of non-fatal cases). To put these estimates in perspective, it has to be remembered that over 70 years about 20% of those who were evacuated (i.e. about 27,000 people) would

67

normally die of cancer.

24. A similar calculation based on the estimated collective doses for the European Soviet Union leads to a number of fatal cancers in the range 5,000 to 20,000, the higher number being based on the upper estimate of collective dose from ingestion of caesium-137. As these values are a small fraction of the spontaneous cancer incidence, the chances of epidemiological detection of the cancers are negligible. In the groups with mean doses substantially above 0.1 Gy one would, however, expect to detect health effects such as leukaemia and thyroid cancer.

FUTURE STUDIES

25. It was clear from the IAEA review meeting that there is a great deal to be learned from the Chernobyl experience about mitigating and assessing the consequences of severe accidents. Accordingly, a number of recommendations were made about international exchanges of information over the coming months and years. The topics to be considered include

(a) design of clothing for protection against high levels of airborne beta activity
(b) assessment, prognosis and treatment of non-stochastic effects in highly exposed people (particularly the acute radiation syndrome and radiation-induced skin lesions)
(c) the methodology for epidemiological studies of selected groups exposed in the Chernobyl accident
(d) the effectiveness of evacuation and sheltering, and the problems associated with introduction of these counter-measures
(e) intervention dose levels and corresponding derived intervention levels
(f) radiological sampling and monitoring under emergency conditions
(g) radiation protection aspects of the decontamination of large areas of land;
(h) use of environmental monitoring data obtained during and after the accident for the validation of mathematical models for radionuclide transfer through the atmosphere, the terrestrial environment and food chains, surface waters (fresh water and sea water), and urban environments.

26. It is anticipated that these subjects and others will be considered in a series of meetings to be arranged by the IAEA and other international organisations. In addition, the United Nations Scientific Committee on the Effects of Atomic Radiation (UNSCEAR) will be assessing the overall radiological consequences of the Chernobyl accident, as part of its continuing assessment of the impact of all sources of radiation.

A remotely organized estimation of the radiation situation in Eastern Europe following the Chernobyl accident

J. R. SIMMONDS, BSc, *National Radiological Protection Board*

Immediately after news of the accidental release from the Chernobyl nuclear reactor became known, the UK Foreign and Commonwealth Office (FCO) asked the National Radiological Protection Board (NRPB) to advise them on the radiological implications. Their primary concern was for British Embassy staff, and for residents in, and visitors to, the USSR and neighbouring Eastern European countries. In this paper the advice that NRPB gave is summarised, the reasons leading to the advice are discussed and the way that the situation changed with time is outlined. In addition, estimates of the radiation exposure in eastern Europe are presented.

THE INITIAL ADVICE
1. In the early period after it became known there had been an accidental release (late April and early May 1986), there was only very limited information on levels of radioactive contamination in the USSR and surrounding countries. Increased levels of radioactivity had been measured in Sweden and from these it was possible to obtain a preliminary estimate of the magnitude of the release. The principal radionuclides detected were ^{131}I, ^{134}Cs and ^{137}Cs, with ^{131}I having the largest concentrations measured. From this initial information we realised that the release was large enough to lead to severe consequences close to Chernobyl and that ^{131}I was the first radionuclide of concern for the contamination of food supplies. Early on, therefore, FCO decided to evacuate students from Kiev and Minsk and to advise people not to travel to the western Soviet Union or to north-east Poland unless absolutely necessary.
2. From theoretical assessments of the radiological consequences of accidental releases (refs. 1-2) and the reported experience following the Windscale fire of 1957 we knew that the foods of immediate concern following an ^{131}I release are milk, together with vegetables and fruit grown in the open. Once ^{131}I is released into the atmosphere some is deposited onto the ground, it is then rapidly transferred along the pasture-cow-milk pathway. Also any crops growing in the open can be directly contaminated by the depositing ^{131}I.

There were also indications that free range eggs could be contaminated depending on the hens' diet.

3.   The areas where levels in food could give rise to concern were not clear in these early stages following the release. However, theoretical studies of potential large accidental releases had shown that restrictions on food supplies might be necessary for hundreds of kilometres from the release point. Indeed restrictions could be required at even greater distances depending on the exact magnitude of the release and the subsequent meteorological conditions.

4.   The appropriate radiological criteria for use in the early phase following an accidental release are the Emergency Reference Levels (ERLs) of dose (ref. 3). For foodstuffs, NRPB has decided that the guidance given by the International Commission on Radiological Protection (ref.4) will form the basis for their advice. Thus, NRPB has recommended that consideration be given to imposing restrictions when projected doses to the public are above 5 mSv to the whole body or 50 mSv to any individual organ, and that restrictions must be implemented if doses approach ten times these levels.
Derived Emergency Reference Levels (DERLs) have been published (ref. 5) giving for example, peak concentrations in foodstuffs corresponding to the recommended reference level of dose for a higher than average consumer.

5.   From the limited information available and taking account of the various factors discussed above, our initial advice to FCO on food and drink was as follows:-

6.   "Travellers to the western USSR, including Moscow and Leningrad, and north-east Poland, including Warsaw, should avoid fresh milk and free range eggs. Surface grown vegetables should preferably be avoided and in any event washed and/or peeled. Fruit should be peeled. It is safe to drink tap-water and to eat meat, poultry and other foods. Anything tinned, frozen or preserved before 26 April is safe. FCO also advised that in all west and east European countries, the advice of the appropriate local authorities on food and drink should be taken into account."

7.   The whole of the western USSR was included in the initial advice because of uncertainties about the scale of the release and subsequent dispersion and deposition of radionuclides. There had been early reports of high external dose rates in north-east Poland, indicating relatively high levels of deposited radioactivity following the release; the advice on food therefore covered this region as well as the USSR. There was no need to give advice regarding foods other than milk, vegetables or fruit at this stage. In many cases this was because the foods would have been produced prior to the release and hence would be uncontaminated. In addition, for some foods, notably meat, levels of radioactivity only build up slowly with time so no precautions were necessary in the short term.

MODIFICATIONS TO THE INITIAL ADVICE

8.   Following the issue of the original advice a number of steps were taken to obtain more information on the levels of radioactivity in the USSR and surrounding countries. We needed this information to enable us to refine the advice and also to indicate what advice might be required in the longer term. Where possible environmental measurements were obtained from the various national authorities. However, some of the early data were contradictory or not reported in a recognisable way and so we realised that this was not going to provide us with enough information. Therefore, with FCO's co-operation, use was made of the British Embassies in Moscow, Warsaw, Budapest, Bucharest, Prague, Sofia and Belgrade. Each Embassy was sent an instrument to measure external γ dose rates and they were asked to collect samples of tap water, fresh milk, vegetables and grass. These samples were sent to NRPB for measurement.

9.   From the various environmental measurements we realised that parts of western USSR were relatively unaffected by the release. For example, as shown in Table 1, the external γ dose rates measured at the British Embassy in Moscow were virtually at background levels. Therefore, the advice on food and drink was withdrawn from these regions, which included Moscow and Leningrad. However, the British Embassy in Bucharest reported relatively high external γ dose rates. There was an indication that this was partly due to heavy, localised rainfall in Bucharest during the passage of the radioactive cloud from Chernobyl. In consultation with FCO the advice on food was therefore extended to Romania. The advice itself remained unchanged, as detailed above in paragraph 6, but it now applied to the Western Ukraine, including Kiev; Belorussia, including Minsk; Lithuania; north east Poland, including Warsaw; and Romania. In addition, restrictions on travel were now only advised for the Western Ukraine and Belorussia.

10.   Over the next week or so, the various food samples were sent back from the Embassies for analysis at NRPB. The levels were compared with the appropriate DERLS (ref. 5), as this was still in the early phase following the release. The radionuclide of concern remained ^{131}I for the initial few weeks following the release. As shown in Table 2, the national authorities in Poland and Hungary reported levels of ^{131}I in milk which approached, or in some cases exceeded 2000 Bq l^{-1}, which is the DERL for the peak concentration in milk corresponding to a dose to the thyroid of 50 mSv for a child drinking about 0.7 l/day of milk. In both countries restrictions on consumption of milk from cattle grazing pasture were introduced. In all cases milk obtained by Embassy staff and measured at NRPB contained levels of ^{131}I well below the DERL. The NRPB measurements on green vegetables were also below the DERL for ^{131}I in all cases.

Table 1   External γ-ray doses

Location	Source of data	Highest measured dose-rate (μSv h^{-1})[1]	Estimated total dose in a year (mSv)
Soviet Union			
Moscow	Brit. Embassy	0.2	(2)
Chernobyl	Official	150	(3)
Yugoslavia	Official	1.6	0.3
	Brit. Embassy	0.7	0.3
Bulgaria	Official	0.7	0.3
	Brit. Embassy	0.3	0.2
Poland			
North-east	Official	4.5	0.8
	Brit. Embassy	1.0	0.6
Warsaw	Brit. Embassy	0.4	0.1
Czechoslovakia			
Prague	Brit. Embassy	0.9	0.2
Hungary	Official	0.4	0.1
	Brit. Embassy	0.4	0.1
Romania	Brit. Embassy	3.6	1

Official figures are from governmental sources in the countries concerned.

Notes:

(1)   No allowance has been made for natural background dose, normally between 0.1 and 0.2 μSv per hour.   These are the highest dose rates known to NRPB, but for some countries, where the first measurements were made about a week after the accident, the measured rates may be less than the peak rates.

(2)   It is likely that the dose-rate given here is due to natural background radiation in Moscow.   However, if this reading is taken to indicate the presence of radionuclides from the Chernobyl reactor then the estimated dose over the year is less than 0.1 mSv.

(3)   It is not clear where in the Chernobyl region these dose-rates were measured.   The dose-rates are likely to vary considerable with distance from the reactor so it is unreasonable to estimate a dose at Chernobyl.

Table 2 Iodine-131 Concentrations in Milk

Location	Peak measured nationally (Bq 1^{-1})[1]	Measured by NRPB (Bq 1^{-1})[2]
Soviet Union Moscow	–	35[3]
Yugoslavia	50 – 150	320
Poland	30 – 2000[4]	560
Czechoslovakia	100 – 500[5]	260
Hungary	200 – 2600	200
Romania	–	420

Notes:

(1) These ranges indicate the variation in the maximum reported concentrations in milk throughout each country. They may not reflect the actual peak concentrations, either because measurements were not made early enough, or because NRPB have not received the appropriate information.

(2) The measurements made by NRPB were on milk samples obtained through shops. They therefore reflect the efficacy of countermeasures as well as the actual peak levels.

(3) The concentration of iodine-131 measured in milk bought in Moscow does not show the usual decay with time; the concentrations are therefore influenced strongly by the source of the milk and countermeasures applied.

(4) These measurements are gross beta; it seems likely that 80–90% of the activity is due to iodine-131.

(5) Concentrations collated by the World Health Organisation rise to 1570 Bq/l iodine-131.

Although the national authorities in Poland and Romania did report relatively high concentrations of ^{131}I in some vegetables, these levels were still below the DERL. Nevertheless the measured levels, combined with a continuing uncertainty in the overall picture of radioactive contamination, meant we did not think it prudent to withdraw any of the advice on food from the countries listed, in the period up to early June 1986.

REVISED ADVICE

11. As discussed above, in the initial few weeks after the release the radionuclide of most concern was ^{131}I. This radionuclide has a radioactive half-life of 8 days and therefore levels of ^{131}I in food would have decayed substantially in six to eight weeks. The radionuclides of concern in the longer term are ^{134}Cs and ^{137}Cs with radioactive half-lives of 2 and 30 years, respectively. In addition the DERLs established by NRPB were no longer the most appropriate radiological criteria for formulating advice on food and drink. As discussed above DERLs are intended for application in the initial phase following an accidental release. For example, in calculating the DERLs no account was taken of the possibility that an individual could consume food such as meat, preserved by processes such as freezing, for lengthy periods. In the longer term such factors have to be taken into account.

12. Although the caesium radionuclides would not have decayed appreciably in the first few weeks following the release, natural weathering processes constantly reduce the levels on grass and other plants. The levels of ^{137}Cs and ^{134}Cs in fresh milk, eggs, vegetables and fruit would therefore be considerably lower after a month or so, than shortly after the release. In contrast caesium levels in meat only build up slowly with time, reaching a peak some 20 or so days after the release.

13. From the available environmental data it was clear that caesium levels in most foods in the period from early June 1986, including much meat, were low enough that the food was acceptable for human consumption, except possibly in the area closest to the Chernobyl reactor. However, it was possible that certain foods in the USSR and other East European countries could have contained high levels of the caesium radionuclides. Milk products, such as UHT milk, powdered milk and cheese, could have been made from highly contaminated milk shortly after the accident. These products could then have been stored to allow the ^{131}I to decay and put on sale; however the caesium would not have decayed substantially in this time. Also milk and vegetables banned from human consumption soon after the accident may have been fed to animals, particularly veal calves, leading to high levels of caesium in the meat. Wild animals, such as deer, consume large amount of mosses and lichens which are known to

concentrate caesium, leading to higher concentrations in meat from game animals than in that from farm animals. Of particular concern was that these foods originating in the area around Chernobyl might have been exported to neighbouring east European countries. However, there were insufficient data to be sure that high caesium levels did not also exist in foods produced in the various countries. Even in areas distant from Chernobyl, where general levels of radiation are of no real concern, localised rainfall may have created small areas with relatively high contamination. In particular, we know that levels of radioactivity in parts of Poland and Romania were significantly higher than in other east European countries. However, in all cases the levels in these foods were thought to be such that only their long term consumption would be of any radiological significance. We do not think that all veal, game meat and milk products were highly contaminated; in fact it was probably only a small fraction of these foods that were affected.

14. The advice to the FCO regarding food and drink in eastern Europe was modified on 12th June 1986 to take account of the change in emphasis from ^{131}I to the caesium radionuclides. The concern now was to prevent long term consumption of possibly contaminated foods and so for the first time a distinction was made between short term visitors (up to about 6 weeks) and residents or longer term visitors. The revised advice to FCO as it was issued by them is given below.

15. "The advice on foodstuffs to visitors to and residents in the USSR is as follows:

(a) Former advice to avoid free-range eggs, fresh milk and non-preserved milk products such as yoghurt, is now withdrawn. But UHT and powdered milk, and preserved milk products such as cheese and butter, which may have been produced locally in May and June should be avoided.

(b) Vegetables and fruit should be washed and, where possible, peeled.

(c) Most meat (e.g. pork, mutton, beef and poultry) is safe to eat. But veal and game animals should be avoided.

Residents in Poland, Romania, Hungary, Czechoslovakia, Yugoslavia, Bulgaria and East Germany are advised to follow similar precautions. Visitors to these countries should follow local advice."

16. Subsequently the advice for the USSR was amended to apply to residents only, as even here the levels of radioactivity in food are likely to be such that only long term consumption should be avoided. The FCO continued to advise people not to visit the Ukraine and Belorussia throughout this time. However, we also issued advice on food and drink in June to people who have to travel to this region despite the FCO advice to the contrary. The advice here differed from the revised advice to the rest of Eastern Europe

in that people were also advised to avoid locally grown vegetables and fruit, lamb and mutton, locally produced fresh milk and yoghurt, and free range eggs if possible. It was pointed out that pork, beef, fresh water fish, and fruit and vegetables grown under glass may have been contaminated depending on local practices. Examples of foods considered safe to eat were also provided. We advised that if possible only local food from official sources should be obtained, as this would have been checked by local authorities. In addition, we pointed out that the amount of radioactivity taken in depends on the levels of the food and the amount you eat, so while eating a little of the foods of concern is all right, people should try not to eat a lot of them.

THE CURRENT SITUATION

17. The revised advice on food and drink issued in the middle of June has remained in force until the time of writing, September 1986. Levels of radioactivity measured by NRPB in samples of food supplied by British Embassies in Moscow and other east European capitals were not high enough to be of concern. In some countries the national authorities also made and reported a substantial number of measurements of caesium radioisotopes in food, and these also indicated that levels were not of concern. However, there still remained the possibility of relatively high concentrations in some foods, either due to particular processes leading to the contamination of the food, or because they could have been exported from the region around Chernobyl. Information on problems of this nature is best obtained from widespread sampling programmes of the type being undertaken by national authorities. The sampling organised swiftly and effectively by the Embassies provided essential information in the early period following the accident, when reliable data from eastern Europe was scarce. However, by mid to late July this sampling programme was no longer necessary and was discontinued in all countries. Similarly, as external γ dose rates fell to background levels it was no longer necessary for the Embassies to report back their dose rate measurements to NRPB.

18. The radiological situation in the USSR was clarified following the International Atomic Energy Agency (IAEA) special meeting on Chernobyl at the end of August. The results of this meeting are reported by Hill (ref. 6) in these proceedings. Also more detailed information has been obtained on radiation levels in various east European countries. As the radiological situation in eastern Europe is now much clearer it is likely that the advice on travel, food and drink will be modified in the near future.

RADIATION DOSES RECEIVED IN EASTERN EUROPE

19. Currently we are carrying out an assessment of the radiation doses that have been or will be received by people resident in the east European countries, excluding the USSR,

due to the Chernobyl accident. We are estimating both individual and collective doses and a distinction is made between exposure in the first year following the accident and that from subsequent years. The radiation doses are being calculated both assuming that no countermeasures were taken to reduce doses, and taking account of any countermeasures. Various countermeasures were taken in the East European countries. For example, in Poland and Romania stable iodine was issued to all children and infants. Restrictions on milk supplies were introduced in all east European countries and in Poland, Hungary and Romania cattle were removed from pasture and given stored, uncontaminated feed. In Hungary people were advised that children should not play in sand and that mothers should wash their children's hands more frequently than usual.

20. Preliminary results of the assessment indicate that, taking account of any countermeasures, the average adult individual committed effective dose equivalent from exposure in the first year following the accident is in the range 100 to 600 µSv. The highest doses are predicted for Romania and north-east Poland, while the lowest predicted doses are for Bulgaria and Yugoslavia. In all cases the doses are lower than those received annually from natural background radiation. (In the United Kingdom the average annual effective dose equivalent from natural background radiation is about 2000 µSv). The doses estimated are also generally within the range predicted in a separate study for the countries of the European Community (ref. 7). Radiation doses several times higher than the average could have been received by some individuals in the population, due to their location in more contaminated areas and because their habits led to their receiving higher than average exposure to radiation. Our results indicate that doses likely to be received in subsequent years are considerably lower than those predicted for the first year. Also, much of the first year's dose is due to exposure in the first few months following the accident.

ACKNOWLEDGEMENT
21. We are very grateful to the British Embassy staff in Moscow, Warsaw, Bucharest, Budapest, Prague, Sofia and Belgrade for the invaluable data that they provided. The work outlined above was carried out by Mary Morrey, Malcolm Crick, Joanne Brown and other colleagues in the Assessments Department at NRPB.

REFERENCES
1. KELLY G.N. and CLARKE R.H. An Assessment of the Radiological Consequences of Releases from Degraded Core Accidents for the Sizewell P.W.R. NRPB-R137. Chilton, NRPB. London, HMSO, 1982.

2. SIMMONDS J.R., HAYWOOD S.M. and LINSLEY G.S. Accidental Releases of Radionuclides: A Preliminary Study of the Consequences of Land Contamination. NRPB-R133. Chilton, Oxon, NRPB. London, HMSO, 1982.

3. Emergency Reference Levels: Criteria for limiting doses to the public in the event of an accidental exposure to radiation. NRPB-ERL2. Chilton, Oxon, NRPB. London, HMSO, 1981.

4. International Commission on Radiological Protection, Protection of the Public in the Event of Major Radiation Accidents: Principles for Planning, Oxford, Pergamon Press, ICRP Publication 40. Ann. ICRP, 14, No. 2, 1984.

5. LINSLEY G.S. et al. Derived Emergency Reference Levels for the introduction of countermeasures in the early to intermediate phases of emergencies involving the release of radioactive materials to atmosphere. NRPB-DL10. Chilton, Oxon, NRPB. London, HMSO, 1986.

6. HILL M.D. Chernobyl - The Radiological Consequences in the USSR. IN Chernobyl a technical appraisal, BNES (These proceedings).

7. MORREY M. et al. A Preliminary Assessment of the Radiological Impact of the Chernobyl Reactor Accident on the Population of the European Community. Commission of the European Communities Report (to be published).

Discussion

PANEL: D. R. SMITH, *National Nuclear Corporation Ltd,*
DR B. EDMONDSON, *Central Electricity Generating Board,*
DR J. H. GITTUS, *Safety and Reliability Directorate, UKAEA,*
H. J. DUNSTER, CB, *National Radiological Protection Board*

Viscount Hanworth, House of Lords
Bearing in mind the difficulty of definition, how much of the
nuclear core inventory escaped at Chernobyl? How would you
answer the layman's question: what would be the effect if, say
60% of the inventory had escaped? Why is this impossible, if
it is the case that it is impossible?

DR J. H. GITTUS, UKAEA
About 3% of the actual weight of the core itself was ejected,
10-15% of the volatile radio-nucleides, i.e. caesium and
iodine, escaped from the core into the atmosphere, about 3% of
the other radio-nucleides escaped and probably 100% of the
noble gases, zenon and krypton.
    It is difficult to say what would have been the effects if
larger amounts had escaped in this accident. For the Sizewell
B pressurized water reactor we considered a whole range of
accidents, ranging up to and beyond the severity of the kind
of accident that happened at Chernobyl. We considered that in
the worst category of accidents, rather than 10-15% of the
volatile nucleides (caesium and iodine) escaping, it could be
60-70%, and in that reactor, where the analysis was very
detailed, we were able to identify processes which resulted in
almost half of the most easily evolved solid elements being
retained in the reactor and not released into the atmosphere.
To date we have not been able to do anything like that kind of
detailed analysis for the Russian reactor system. This kind
of risk assessment coupled with calculations of how much might
be released and what kind of accidents might lead to those
releases is not something that can be done overnight, nor
without spending a great deal of money. It runs into hundreds
of thousands of pounds to do this kind of assessment and
therefore we tend only to do it for our own domestic reactors
that we propose to build. But when the analysis was done
there would not be much more than half of the most volatile
species evolved. Of course, in this particular case the
Russians were able to intervene; they improvized a sand filter
on top of the reactor, and undoubtedly that has had something
to do with the fact that 85% or 90% of the caesium and iodine
did not escape.

As far as the late effects are concerned if more had escaped, these are assumed proportionate to the societal dose via the constant of proportionality, which is approximately 100 sieverts per late cancer, and the physics is such that if all that has happened is that there is, say, twice as much released, and everything else remains the same, then approximately twice the societal dose would be expected, and thereby through the linear hypothesis, twice the number of late effects.

There is also a possibility of early deaths, not just among the firemen who intervened to stop the fire from spreading, but also among members of the public living close to the reactor. In a number of the accidents which we postulated and analysed for Sizewell B, there could be early deaths among that community. The severest accident, which was called UK 1, had a calculated frequency of about three times in 100 million years of reactor operation – and this is a measure of the reliability of that particular plant. The expectation for average weather conditions, from National Radiological Protection Board calculations based on authority source terms, was 130 early deaths among members of the public.

Therefore in a more severe accident – and another accident with a reactor of any kind could be envisaged – then there could be not only a proportionate increase in the number of late deaths but also the emergence of a small number of early deaths among members of the public as well as emergency workers.

The thing that always has to be said when talking about effects is how likely it is that an accident of this kind will occur.

Surgeon Captain J. R. Harrison, Royal Navy

Is there a possibility that the Chernobyl public safety scheme advised early shelter and delayed evacuation? Do the panel think that early, automatic, pre-planned evacuation is potentially dangerous?

MR H. J. DUNSTER, NRPB

Yes; I think evacuation is always slightly dangerous. I think that at Chernobyl they did not evacuate Pripyat on day one for two reasons. Firstly, I do not think they were ready for it. They had to get their thousand buses together. Secondly, they had a deposition to the side of Pripyat rather than over Pripyat on the first day, and there were disadvantages in moving people through that deposit so they sheltered them and gave them stable iodine and evacuated them about 24 hours after the start of the event. I think that is probably quite sensible.

I do not think an action like automatic evacutation on an emergency plan should ever be taken. I think it should be a decision taken by the emergency controller in the light of the circumstances. Certainly an emergency plan has to provide for it if the emergency controller so desires.

Mr P. Nalpanis, Technica Ltd
What lessons can be learnt from Chernobyl regarding the
modelling of atmospheric dispersion and subsequent ground
imposition of radio-nucleides?

MR DUNSTER
I think it is yet another bit of practical data to feed in.
Generally, although straight line models look nice on paper,
they do not give the map actually expected on the ground.
Even the first day's discharge looped its way round Eastern
Europe before arriving in the UK, and the only way this can be
dealt with is as a complicated meteorological forecasting
exercise. The links with the Meteorolgical Office on any
occasion like this are absolutely vital.

Lieutenant Commander W. Nimmo-Scott, Royal Navy
At what level was the experiment on the reactor authorized,
and how good were the procedures?

DR B. EDMONDSON, CEGB
The initiation of the test was in the hands of the electro-
technical experts in the Moscow Institute, as was the
preparation of the procedures. This was declared by the
Soviet experts to have been badly done. Certainly there was
no real evaluation of the safety significance of the test
before authority to proceed was given. This authority, in the
final analysis, was in the hands of the station management, in
the sense that the safety of the plant is the responsibility
of the station manager. It was up to him to make sure that
the test procedures were satisfactory and that the proper
safety analysis had been done, and that he had received a
formal statement about a safety analysis done with competence.
That is the arrangement and it was not properly handled,
because both the test procedures and the safety analysis were
inadequate.

Mr J. Dickson, The University of Manchester
It would appear to have been remarkably easy for the operators
to by-pass reactor trip CCCS, activation, etc. Is this
correct?

DR EDMONDSON
The two-turbine trip veto was straightforward - by means of a
single switch on the operator's desk. Occasionally there is a
need to veto trips in the normal operation of the reactor, for
example, during start up and shut down. This veto was an item
which had to be available in the operator's hands. The answer
to whether a veto could be made more difficult is yes, but the
requirement for the procedure is fairly regular. As to the
vetoing of the other trips, I am not quite sure, but I think
it is fair to say that it must have been quite easy as far as
the drum level and steam pressure trips were concerned because
the action was taken quite quickly at a late stage in events.

Clearly it was not necessary to send a man to get a manual and screwdriver and necessary keys.

Mr E. G. Bridges, CEGB

In view of the comment that the operators were the key to the accident, would the panel comment further on the academic qualifications, training and experience for staff in the control room at the time of the accident?

DR EDMONDSON

I have no knowledge of that. One of the things that I hoped was emphasized was that while it is true to say that the operator's actions were one key to the accident, I hope at least equal weight was placed on the design features which magnified the operator's actions to disastrous effect, and which were inadequate to prevent him pursuing the gross mis-management of the plant.

Mr D. Whitfield, Nuclear Installations Inspectorate

What information is available on the respective structures of the interactions between operating and testing teams, and the length of experience of RBMK operation and the members of the operating team?

DR EDMONDSON

I cannot give you any information on that, and I do not think my colleagues would be able to either. It is not a topic which we entered into, largely because we were concerned not so much with the practices anad procedures of that kind but just what happened and why it did happen as a result of the actions that were taken. We did not go into the management side of things in any great depth.

DR GITTUS

The agency has set up a working group. The idea is that the Board of Governors of the agency will receive recommendations from the secretariat which have been formulated by this working group concerning the scope and extent of a programme of future international collaboration in the remainder of the financial year and the year after, together with how much it is going to cost and how the money is to be accumulated. At the moment there are 13 topics to my knowledge that may be the subject of collaboration, and some of these are to do with operators. The first topic is severe accident analysis. The second is the man-machine interaction - the balance between automatic control and manual control of nuclear reactors. The Russians, for example, supplied their operators with a computer which towards the beginning of the accident gave out the message: reactor is in an unsafe state; it must be closed down. The operators did not trip it. Had they done so, the accident might not have happened. The Russians implied that it was really traditional that this particular trip was left to the operators, and with hindsight they realized that they

should have automated it. They had not got the balance
between automatic control and manual control right in that
particular area. How to strike that balance is another topic
for future discussion.

The next topic was operator error and then operator
training. There are three topics here which are squarely in
the area addressed by a number of questioners, and I hope that
this interaction will cover not only plans for the future,
i.e. how to train operators for the future, what balance to
strike for the future between automation and manual control,
and how to record operator errors in a way which will enable
us to learn lessons in future, but will also be retrospective
and answer questions about how well-trained the Russian
operators were, and what their qualifications were.

Other areas to be discussed were applications of the lessons
learned, the derived emergency reference levels in different
countries, decontamination, radiation monitoring, dosimetry,
epidemiology, intervention to reduce early effects, e.g.
surgery and bone marrow transplants, and intervention such as
the giving of potassium iodate to minimize late effects.

Mr A. C. Hall, Westinghouse Nuclear International
Which regulations or practices applied in the UK, had they
been applied at Chernobyl, would have either prevented the
accident or rendered its occurrence much less probable?

MR DUNSTER
One word, I think - licensing. It seems to me that one of the
weaknesses of the system which I understood was disclosed by
the discussions in Vienna was that there had not been an
adequate second or even third look at the design and the
interrelation between design and operator errors. The
statement was made quite openly that the designers had never
believed that operators would act so stupidly. It is one of
the fundamental tenets of the licensing organization that it
expects operators to behave stupidly. It is relieved when
they do not, but it does assume that they will sometimes, and
I think that must be the case.

The point about licensing is not so much that it is a legal
process but that it is a second and quite separate look. The
designer does not want to have accidents - it is bad for his
reputation if nothing else - so he has been quite careful, but
someone else who is not thinking along the designer's lines is
needed to go from first principles and say that it is not
sensible to assume that no one will make mistakes.

THE CHAIRMAN
And is it clear that there was no Russian equivalent?

MR DUNSTER
Not clear in my view, but I do not see it on the ground.

MR D. R. SMITH, NNC

The safety approach in this country is to try to identify
which operator actions could conceivably lead to serious
accidents, and to design into the engineered plant protection
to prevent the accident happening, irrespective of operator
action. This can be done by interlocks which physically
prevent the operator action taken from having the undesired
effect, or it can take the form of a reactor protection system
which recognizes that there is a transient developing, and
automatically shuts down the reactor by tripping the control
rods. We obviously rely on operators behaving in a rational
way, but we try, successfully I think, to ensure that even if
operators do not behave rationally then the protection systems
on the reactor will prevent the situation escalating in the
way it did at Chernobyl.

Mr J. R. W. Marsh, Royal Naval College

Before the accident, Chernobyl was being operated outside the
design intent. Operating documentation and operator training
concentrates on operation within the design intent. Should
operators be trained to understand the consequences of
operation outside the design intent?

DR EDMONDSON

It is self-evident that the Chernobyl type of accident is not
possible in the UK because of the specific design features of
the plant. I think this is an important point to make because
many of the things that have been said have emphasized the
part played by the operator and made less of the aspect of
design shortcomings. In fact it was recognized ten years ago
that the RBMK design of the plant would not be able to be
licensed in the UK.

At the time of the accident the operation of the plant was
what one might loosely call off-normal. A system test was
being carried out where the plant was put into a special
configuration and run in an abnormal fashion. It is clear
from what has been said that there was no proper safety review
of that process. In the UK we have excruciating safety
reviews of off-normal practices. As chairman of the body in
the CEGB that from time to time has to oversee these reviews,
I get many complaints about the grave difficulty and the time-
consuming work that is involved in obtaining approval for off-
normal practices on any of our plants. The fullest possible
off-site safety review is necessary before a station manager
can embark on anything which is outside his normal operational
procedures.

Perhaps the single lesson that strikes me most about the
Chernobyl accident is that the operators cut corners by
vetoing safety systems and ignoring warnings. I believe they
could only have done that in total lack of perception of the
consequences of what they were doing. They seem to lack a
broad education whereas operators in the UK have a well-
founded basis to their training. I believe there are three

separate practices that would prevent the Chernobyl accident occurring here: plant with the design features which led to the accident would not have been licensed in the UK; extreme care is taken in the safety approval of off-normal procedures in the UK nuclear power stations; and the strength of the educational aspects of operator training in the UK.

Mr S. P. Rao, Impell Corporation

In the light of Chernobyl, what does the panel think about the low population zone limit and the relative changes in emergency procedures in the UK?

MR DUNSTER

I do not think that one changes emergency procedures on the basis of someone else's accident, unless one of two things has happened - either that the accident sequence has drawn attention to a situation not previously taken into account, which I do not believe to be the case with this particular accident, or that the mechanism of handling the accident was in some way a lesson. Then I think there is something to be said. There is always something to learn from somebody else's practical accident. The planning of emergency action in the event of an accident is a somewhat unrealistic process, inevitably, because although exercises are run, they are not as realistic as the real thing in the end. When somebody actually has an accident, what they did can be seen, and some real lessons can be learnt. I think it is too soon to say what those lessons are. Perhaps one of the key points is that learning lessons of this kind ought to be regarded as a matter of years not months. There is no point in rushing off, deciding what ought to be done, and trying to do it quickly. This is not an urgent proposition, it is something which is talked through over a period of at least two years, in my view.

Mr Y. Capouet, European Commission

Western safety requires an absolute separation of safety system with electricity-producing equipment. The RBMK's safety approach, however, permits the use of the turbo-generators as a safety electrical source. Is there any move on the Russian side to review this fundamental choice, and are there any technical possibilities that will enable them to do so?

MR SMITH

I do not think it is true to say that Western safety requirements preclude the use of the turbo-generator or any other normally running piece of equipment for safety purposes. They do require an assessment to be made of the reliability of that equipment, and therefore of the likelihood that it would be available when required to perform the safety function. We also require diverse protection for any frequent fault in the UK, i.e. any fault that has a likelihood of occurring more than once in a thousand years, and that obviously means that there has to be some other piece of equipment other than the normally operating equipment capable of performing the safety function. In the example taken, it could be the emergency diesels, the batteries, or some non-electrical prime mover that could provide the safety function when required.

I do not think therefore that there is a difference in principle here, but there is a difference between the extent of redundancy and diversity which we provide and that provided by the Russians. However, since 1975 when we first looked at the RBMK, the Russians have moved quite some way towards such provisions, but they have a long way to go before they reach the levels required in the UK.

Mr G. Fordyce, CEGB

You have concentrated on the effects of the release on Eastern Europe. Could you briefly outline the effect on the UK and explain why restrictions on lamb sales are still in force?

MR DUNSTER

The deposit in the south of England was small. The deposit in the heavy rainfall areas was a good deal higher. Broadly speaking, the average radiation dose across the whole country was about 50 micro-sieverts over all time for all routes of exposure, compared with a few thousand micro-sieverts per year from natural sources. The figures in the wet areas were maybe of an order of magnitude higher than that.

I do not know enough about the Ministry of Agriculture's measurements on lamb to know why the ban has not yet been lifted. The figure they adopted was derived largely from the figure which was being used in Europe at the time rather than one which was worked out from first principles, and it was difficult to use a higher figure. It did assume that lamb taken from the highest area at the highest time would be put into a deep freeze as a form of staple item of diet for somebody, so it was a fairly cautious assumption. The actual dosage delivered by eating lamb appeared to be very much lower than was thought at the time. But the number exists and has been maintained, and is a perfectly sensible programme for consistency. I do not know what the results in lamb show at the moment, and I have no idea for how long the restriction is going to persist, but probably not much longer.

Mr I. J. G. Berry, Middlesex Polytechnic

Bearing in mind that the positive void coefficient was a major factor in the Chernobyl accident, do members of the panel see any significant imputation regarding the future of LMFBRs?

MR SMITH

Simplistic statements to the effect that positive void coefficients are unacceptable should be avoided. Canadian reactors have a positive void coefficient, but they are not boiling reactors so there is no significant coolant voiding during operating transient. The same applies to the LMFBR, which is a sodium-cooled fast reactor. A reactor would have an increase in reactivity if the sodium was voided from the core, but in this case the maximum coolant temperature at full power is some 300°C below the boiling point of sodium, so voiding would only occur in an extremely severe transient resulting from complete failure of the reactor protection system.

To draw a simplistic dividing line, it is best not to have a positive void coefficient in those types of reactors in which there is a bulk boiling in the core, i.e. direct cycle reactors. It is a matter that has to be looked at in detail in the safety behaviour of the particular reactor type.

Dr A. J. Wickham, CEGB

There have been conflicting statements made about high graphite temperatures together with an inherent RBMK problem of high stored energy in the graphite. With early plant of this design known to suffer excessive graphite neutron damage, could a summary be made of the known physical, mechanical and chemical properties of the moderator and the way in which these might have contributed to the subsequent behaviour?

MR SMITH

Not much is known about the nature of the graphite used by the Russians. We do know from our own experimental work that the behaviour of graphite under irradiation depends quite critically on the raw material and manufacturing route. The Russians may have data which would support the behaviour of graphite in their context, but we certainly have not seen it. We would not expect Wigner stored energy to be significant at the graphite temperature in the RBMK reactor.

DR EDMONDSON

It may be true that the graphite fire introduced a larger, more extended release than might otherwise have been the case. However, I think that the graphite in the Chernobyl accident is very much a secondary issue in the sense that it certainly did not contribute to the initiation of the accident. If it played any part at all it was in a relatively late stage, in the consequential aspects rather than in the initiation aspects.

MR SMITH

I agree, but there are other kinds of accidents than the one that happened at Chernobyl. An accident which did not start as a reactivity power surge, but as a rupture for some reason of the membrane which surrounds the graphite stack, is easily conceivable. If that happened, allowing air ingress to graphite at a temperature of 700°C or thereabouts, then that graphite would ignite and possibly cause the fuel channels to fail. So that although the graphite did not have a major influence on the actual accident that happened at Chernobyl,

it is a feature of the RBMK reactor which could be significant
in other accident scenarios.

Mr D. G. Arnott, Rowntree House Associates

The analysis has been fascinating, but it is all physics.
Where is the chemistry? Is there to be an analysis of the
progress of the fire, without which there would have been no
European-sized problem? There are obvious implications for
core design and materials choices such as Zircaloy in
conjunction with steam.

MR DUNSTER

The first part of the question is a misjudgement. I think the
initial release was quite independent of the subsequent
graphite fire. Maybe half the release was from the graphite,
but the initial release in itself was enough to have a
substantial effect on Europe.

MR SMITH

In the second part of the question, the questioner is thinking
of zirconium-steam as being a disadvantage, but he may be
thinking that stainless steel and carbon dioxide do not
exhibit a vigorous chemical reaction. Obviously that is
something to be taken into account. The possibility of
vigorous chemical reaction with the coolant fuel clad is a
disadvantage safety-wise, but provided that the reactor design
ensures that the fuel temperature is kept below specified
limits, it is acceptable.

DR EDMONDSON

I think that the game was lost as a result of explosive
generation of steam from the major input of heat arising from
the reactivity transient. The big kick came from the physical
interaction between the disintegrating fuel and water.
Although the water-Zircaloy reaction and the graphite-air
reaction might have had a part to play, in my view the bulk of
the problem was produced by a physical transfer of heat to the
coolant. After that the game was lost. The chemistry comes
into play at some stage, but not as anything to do with the
initiation, or with major differences in consequences.

DR GITTUS

I would support that. The accident seems to have been one in
which there was a very large deposition of energy as a result
of a prompt criticality. This was thermal energy initially,
but it was transformed into mechanical energy by the
generation of steam, probably explosively, and that is the
cornerstone of the accident sequence. Although there were
other things which no doubt added to the damage, such as the
zirconium-steam reaction with hydrogen production and the
reaction of graphite with steam and air to produce carbon
monoxide and hydrogen (all of these inflammable gases), I do
not think that they were significant in this context. Whether

or not these things happened, given the cornerstone, the accident would have had these dramatic effects and have produced this quite large release of activity.

Professor G. N. Walton, Imperial College and Watt Committee on Energy Ltd

Would it be possible to design and cost a PWR with endo-thermic reactions? Why do we not have steel cladding in the cost of the PWR?

MR SMITH

The simple answer is that there are many aspects to be taken into account in designing a reactor. Stainless steel cladding requires higher enriched fuel and is more susceptible to failure due to stress corrosion in water. Zirconium cladding has been adopted for most water cooled reactors, and the safety characteristics have been shown to be acceptable.

DR GITTUS

In the safety analysis of our reactors we always take account of accidents in which there is what is called a reactivity insertion. Reactivity insertion accidents from various causes are included in the analysis. I think it is very easy, particularly for members of the public, to believe that something has happened for which there was no precedent and which had been completely overlooked in the safety analysis of our reactors. That is not the case. We have not learnt lessons about Zircaloy corrosion, hydrogen production, prompt criticality, steam explosions, or graphite fires. The fact that they can occur is well known to safety specialists , in Russia too, and in our reactors when we do the safety analysis we envisage accidents of this kind which embody phenomena of that type. When we convince ourselves and the licensing authorities at the end of those analyses that the reactor is safe, it is safe in those terms. There is therefore no reason to rush off and suddenly change something that figured in the Russian accident sequence.

Mr B. R. Cundill, GEC Energy Systems Ltd

Would the panel comment on the hypothesis of Dr Ospina that diverse fall-out patterns suggest a failure of the spent fuel storage pool about a week after the main accident?

DR EDMONDSON

I do not know how those patterns would be interpreted, but certainly the question of what happened to the spent storage pool was addressed to our Russian colleagues and it was made very clear that it was intact, that there had been no releases from it and that there was very little fuel in it anyway.

Dr S. J. Strachan, Associated Nuclear Services

How have the emergency monitoring schemes within the Soviet Union coped with the effects of air and ground contamination from Chernobyl compared with the response of the US and UK schemes to the accidents at Three Mile Island in 1979 and at Windscale in 1957? What conclusions can be drawn from such comparisons for the resources required for emergency monitoring of design-based accidents?

MR DUNSTER

I think the lessons learnt are basically that a better system of central collection of data from more monitoring stations is needed. These things cannot be done separately. The NRPB has an existing responsibility under the emergency schemes to encourage, carry out and collate the data from monitoring schemes outside the immediate area of impact of the accident – so-called information monitoring. Some of the gaps between existing nuclear installations could be filled in, and I am sure that will happen. There could be improved methods of communication of that information to the centre and much better methods of communicating it back to the people providing it. These are quite complicated things and it is not going to be done easily. It does not just mean putting in telephone lines. It means looking at our data-handling practices in their entirety, which is one of the reasons why I say I do not want to rush it.

A great deal has been learnt from the difficulties of handling this accident. The Russians appear to have done very well. Their system was simply to measure a dose rate one metre above ground from gamma radiation once the cloud had gone by, and then model the effect of milk in children, caesium in foodstuffs, direct radiation from the ground, doses in houses, etc. They appear to have got it broadly right; by having single gamma ray dose rate measurements scattered across the country they were able to predict all these other consequences with a reasonable precision. I think it was within a factor of two or three, because there would have been casualties. I think we have something to learn and a lot to put on the ground in the form of hardware.

Dr J. D. Lewins, The University of Cambridge

How far does the Chernobyl accident support the popular view that scientists in white coats design and engineers in overalls have accidents? Should the engineer be subservient to the scientist? Most research scientists acknowledge that any test or experiment of the contact system requires at least some operating rules to be suspended. Is this inevitable?

MR K. J. SIMM, INuclE

We have sometimes come across obstructions from the CEGB who thought we were going too fast and did not want to get too involved in computer control systems. I think it is unfair of scientists to think of engineers as always wearing blue overalls. We do occasionally do a little bit of designing. This is not the forum to go into the definition of what an engineer is. But I do feel that we have progressed a little way and we do have some experience on our side.

As far as the operation of plant is concerned, some operating shortcuts have to be taken into account, but these are very well supervised indeed.

DR EDMONDSON

I think that Dr Lewins is absolutely right; there are always going to be procedures where one has to set aside what could be described as an operating rule. There is nothing wrong with this, providing it is done in a controlled and proper fashion. One has only to think of plant maintenance procedures. If for example an operating rule requires that two pumps are available and it is necessary to take one out of service for a brief period to carry out maintenance, there is nothing wrong in suspending that rule to make sure that the plant is properly maintained, providing it is done under a proper discipline and compensating arrangements are made. Under proper control this is perfectly satisfactory, and it is what is done in practice.

MR SMITH

I challenge Dr Lewins' proposition. I have worked in the nuclear industry as an engineer for thirty years and I have never felt myself subservient to scientists.

Mr J. A. Dukes, INuclE

There will some day be an internationally recognized scale of magnitude for nuclear accidents analogous to the Richter Scale for earthquakes. Does the panel think that such a scale is best restricted to civil nuclear power generating systems, or ought it to be so devised as to be able to measure the generality of nuclear mishaps?

DR EDMONDSON

I know that there is a good deal of thinking going on in this field, but I am not sure that I can say precisely what it is because the whole matter is in embryo. Certainly the idea of an equivalent to the Richter Scale to describe nuclear accidents in terms of their severity is something much on people's minds. Maybe Mr Dukes has something on his mind on this topic that we ought to know about as well.

MR DUNSTER

I think it is wicked and stupid because it sets out consciously and deliberately to conceal things. A nuclear accident is never going to be a nice simple event which leads to one number. To do so is to deliberately attempt to either soothe people or simplify it to the point of concealment. It is a fundamentally false objective.

DR EDMONDSON

I do not agree. I think that simply because of the complexity of the issue with which we are dealing, a shorthand has to be found. We do not have a shorthand and that is why we get ourselves into such grave difficulties. Nobody could describe an earthquake as a simple matter and certainly seismic experts do not, but we understand what 5.7 on the Richter Scale means. Nobody knowledgeable would describe a seismic event as in any way simple.

MR DUNSTER
The Richter Scale is simple.

DR EDMONDSON
Yes, but your criticism is that the event to be described is not simple, and therefore a simple scale is unsatisfactory.

MR DUNSTER
No, the number is satisfactory only because it describes one single quantity, and it is possible to relate that to a reactor accident by saying it released so much proportion of the fuel or so much proportion of the energy, but it is not relevant in the case of a reactor; it is reasonably relevant in the case of an earthquake. I think it is a dishonest approach.